D0712833

THE COMPREHENSIBLE COSMOS

Where Do the Laws of Physics
Come From?

—VICTOR J. STENGER—

THE COMPREHENSIBLE COSMOS

COSMOS

Where Do the Laws of Physics Come From?

The most incomprehensible thing about the world is that it is comprehensible.
—Albert Einstein

Prometheus Books
59 John Glenn Drive
Amherst, New York 14228-2197

Published 2006 by Prometheus Books

The Comprehensible Cosmos: Where Do the Laws of Physics Come From? Copyright ©
2006 by Victor J. Stenger. All rights reserved. No part of this publication may be reproduced, stored in a retrieval system, or transmitted in any form or by any means, digital,
electronic, mechanical, photocopying, recording, or otherwise, or conveyed via the
Internet or a Web site without prior written permission of the publisher, except in the case
of brief quotations embodied in critical articles and reviews.

Inquiries should be addressed to
Prometheus Books
59 John Glenn Drive
Amherst, New York 14228–2197
VOICE: 716–691–0133, ext. 207
FAX: 716–564–2711
WWW.PROMETHEUSBOOKS.COM

10 09 08 07 06   5 4 3 2 1

Library of Congress Cataloging-in-Publication Data

Stenger, Victor J., 1935–
    The comprehensible cosmos : where do the laws of physics come from? / Victor J.
Stenger.
        p. cm.
    Includes bibliographical references and index.
    ISBN 13: 978–1–59102–424–8 (hardcover : alk. paper)
    ISBN 10: 1–59102–424–2 (hardcover : alk. paper)
    1. Physics—Philosophy. 2. Physical laws. I. Title.

QC6.S8119 2006
530.01—dc22   2006009274

Printed in the United States on acid-free paper

CONTENTS

PREFACE

In a series of remarkable developments in the twentieth century and continuing into the twenty-first, elementary particle physicists, astronomers, and cosmologists have removed much of the mystery that surrounds our understanding of the physical Universe. They have found that the cosmos is, on the whole, comprehensible. Or, more precisely, they have mathematical models that are currently consistent with all observational data, including measurements of incredible precision, and they have a good understanding of why those models look the way they do. While these models will undoubtedly be superseded by better models as science continues to advance, the great success of current schemes makes it likely that they are on the right track. The broad picture that is drawn by modern particle physics and cosmology describes very well what we see in nature. What we have yet to learn may be expected to fit comfortably on its foundation, just as these sciences fit comfortably on the foundation of centuries-old Newtonian physics.

In one of the most influential books on the philosophy and history of science in the twentieth century, *The Structure of Scientific Revolutions*, physicist-philosopher Thomas Kuhn argued that science does not progress gradually but instead by a series of revolutionary "paradigm shifts."[1] However, in a retrospective published a generation later, Nobel laureate physicist Steven Weinberg concluded that there have been very few true scientific revolutions besides the Newtonian.[2] While, of course, we cannot rule out any future dramatic developments in human thinking, at this writing no observation or theoretical problem forces us to conclude that a major paradigm shift will be required for us to advance further. This is not to say we have reached "the end of science," just that we have reached a plateau where we can see well in the distance where science is going. Before we get too far, let me state explicitly that I am not saying that science knows everything, but I am claiming that science knows a lot more than most people, even many scientists, realize.

The models of physics are commonly assumed to represent lawful aspects of reality. In the minds of scientists and laypeople alike, the statements contained in these models, conventionally referred to as "laws," are likened to the civil and perhaps moral laws that govern human behavior. That is, the laws of physics are thought to represent a set of restraints on the behavior of matter that are built into the structure of the Universe, by God or by some other all-abiding governing principle. I believe many physicists would agree with the view expressed by our highly respected colleague, Kip Thorne, who wrote in his best seller *Black Holes & Time Warps*,

> Throughout this book, I shall adopt, without apology, the view that there *does* exist an ultimate set of physical laws (which we do not as yet know but which might be quantum gravity), and that those laws truly *do* govern the Universe around us, everywhere. They *force* the Universe to behave the way it does.[3]

This is essentially the conventional view. However, as philosopher of science Bede Rundle recently put it, "It is high time this convention was abandoned."[4] In this book I abandon that convention. I will show that the laws of physics are simply restrictions on the ways physicists may draw the models they use to represent the behavior of matter. These models describe scientific observations of an objective reality that surely exists independent of human thoughts. However, their specific forms depend very much on those thoughts. When this fact is recognized, many of the so-called mysteries surrounding modern physics melt away.

Now, I know from commentaries I have already received that this notion is not easily understood and, indeed, is usually gravely misunderstood. I am not claiming that space, time, matter, or any of the other elements of our models do not exist "in reality." However, I cannot prove that they do; furthermore, I have no need to provide such a proof. As long as our models successfully describe previous observations and predict future ones, then we can use them without ever getting into metaphysical questions. This is a point you will find me making continually throughout this book.

Again, my philosophical differences with Kip Thorne cannot be starker, despite the fact that I do not have a single reason to dispute any of his physics claims. As Thorne further elaborates his position,

> This viewpoint [Thorne's] is incompatible with the common view that physicists work with *theories* which try to describe the Universe, but which are only human inventions and have no real power over the Universe. The word *theory*, in fact, is so ladened with connotations of tentativeness and human quirkiness that I shall avoid using it whenever possible. In its place I shall use the phrase *physical law* with its firm connotation of truly ruling the Universe, that is, truly forcing the Universe to behave as it does.[5]

I share Thorne's aversion to the word *theory*, but for the entirely opposite reason. While the term is often misused by laypeople and associated with speculation that is not to be taken seriously, as in "evolution is only a theory," scientists and the science-savvy assign a much greater significance to the term. Few people think of the "theory of gravity" or the "theory of relativity" as "ladened with connotations of tentativeness and human quirkiness," though, in fact, they are. So while Thorne prefers to call a *theory* a *physical law*, I prefer to call it a *model*. Indeed, this has become a more frequently used designation, with the highly successful and hardly speculative "standard models" of physics and cosmology.

So, I shall try to avoid using the word *theory* as well, only replacing it with *model* rather than *physical law*. This will not always be possible since the word *theory* is included in the familiar names of many models, such as "Einstein's theory of relativity." For a similar reason, I will not drop the use of the term *law* since that is how scientists and laypeople continue to refer to the basic principles of physics. But while Thorne chooses to drop any qualifiers between what he calls "true laws and our approximations to them," I will try to make it clear that the laws of physics are the way they are because they have been defined to be that way.

Physics is formulated in such a way to assure, as best as possible, that it not depend on any particular point of view or *reference frame*. This helps to make possible, but does not guarantee, that physical models faithfully describe an objective reality, whatever that may be. This is not to say that objective observations cannot be made from one particular point of view. They can. This can be accomplished by having several independent observers in the same frame. However, models that vary depending on a reference frame cannot be expected to have the universal value of those formulated to be frame independent.

In this regard, let me further clarify my use of certain terms. I will use both *reference frame* and *coordinate system* to refer to any particular point of view, whether within a framework of space and time or the more abstract mathematical spaces we use to formulate our models. Generally speaking, a coordinate system is more specific in terms of providing a framework for a record of any measurements made in that reference frame. An example of a space-time reference frame might be where I am now sitting, at my computer terminal. A corresponding coordinate system might be a set of Cartesian coordinate axes fixed to Earth with its origin in Lafayette, Colorado, and three coordinate axes x, y, and z, where z points vertically upward, x points to the North Pole, and y points west along the latitude line passing through that location. A different point of view would be obtained from a reference frame on the moon, where measurements could be referred to a coordinate system located some-where on the surface.

In later discussions, I will use a rather simple example of a reference frame or coordinate system that is not based on familiar spatial measure-ments but on more abstract representations of data. Imagine an economist who specifies the state of the economy as a point on a graph in which one coordinate axis is *supply* and another *demand*. A different reference frame or coordinate system would be that of another economist who prefers to specify the state of the economy as a point on a graph in which one coor-dinate axis is *wages* and another *profits*. These are two examples of abstract spaces. Of course, economies are more complex, but this should serve as a useful visual aid to help understand the common use of abstract spaces in physics.

When we insist that our models be the same for all points of view, then the most important laws of physics, as we know them, appear natu-rally. The great conservation principles of energy and momentum (linear and angular) are required in any model that is based on space and time, formulated to be independent of the specific coordinate system used to represent a given set of data. Other conservation principles arise when we introduce additional, more abstract dimensions. The dynamical forces that account for the interactions between bodies will be seen as theoret-ical constructs introduced into the theory to preserve that theory's inde-pendence of point of view.

I do not want to leave the impression that every facet of the Universe is thereby "explained" or, indeed, ever will be. In this regard, "explanation" is taken to mean a statement accounting for all the details on something, while "understanding" or "comprehension" is taken to mean the perception of the nature of something. It is in this sense that I claim the cosmos is comprehensible.

The hope of any ultimate, complete theory from which every detail of the world may someday be derived is actually dimmed by the new developments. Instead, many of the structural details of the Universe, including basic facts like particle masses and force strengths, may turn out to be accidents. If so, the origin of cosmic structure can be likened to the origin of biological structure, the combined result of tautological necessity, self-organization, random chance, and perhaps even some natural selection.

These assertions may come as a surprise to many readers, even those with strong scientific backgrounds. We continually read breathless reports in the popular media about the enigmatic new worlds being uncovered by contemporary science. And we read speculations about an equally obscure, ultimate "theory of everything." But mystery can merely mean unfamiliarity; it need not imply lack of understanding. The so-called mysteries that currently exist can be understood, if not explained. The cosmos is comprehensible.

A reader of this book may occasionally get the impression that I am promoting a *postmodern* doctrine, which asserts that science is just another cultural narrative and that one theory is as good as another. Also, at times, I may be seen as arguing that humans make their own reality— a common claim we often hear these days to support the notion that quantum mechanics provides a mechanism for imagined paranormal phenomena. Neither of these describes my position.

While the models of physics might be thought of as cultural narratives, not all cultural narratives are equivalent. Science is by far the most successful "cultural narrative," and none of the others even come close to matching its success. You might think of science as a 100 megapixel digital camera taking pictures of whatever reality is out there, compared to drawing a picture in the dirt with a stick. The scientific picture gives a far better representation of that reality, but we still must be careful not to equate the image with reality.

The notion that the mind can make its own reality is a gross misrepresentation of quantum mechanics, which I have dealt with in detail elsewhere and is only taken seriously in the pseudoscientific literature of parapsychology and alternative medicine.[7]

I also need to make clear upfront that I am not proposing any new physics. Rather, I am trying to show that the existing well-established models of physics can be understood more simply than most people imagine and that they are not filled with unfathomable esoterica accessible only to an anointed few.

This is not to say that science is easy. I personally worked for forty years at the forefront of research in elementary particle physics and astrophysics and can testify that no new discovery comes without effort. Science is often excruciatingly hard work, and not always immediately rewarding. More of the projects I was involved with were failures or did not live up to expectations than were resounding successes.

Nevertheless, I played a small part in a number of major discoveries and witnessed many others close up. I was a collaborator on experiments at most of the major particle physics laboratories in the world. Additionally, I have made astronomical and cosmic ray observations on mountaintops and in mines, and helped pioneer the use of the deep ocean as a physics laboratory. It is the perspective gained from that experience which I bring to this book. The viewpoint I present will be that of a strict empiricist who knows of no way to obtain knowledge of the world other than by observation and experimentation.

Unfortunately, current physical understanding is often formulated in terms of higher mathematics that is as indecipherable to the layperson as hieroglyphs in an Egyptian tomb. No matter how valiantly scientists and science writers may work to express modern models in the vernacular, precise mathematical descriptions and their associated logical deductions can never be satisfactorily reduced to everyday language. Describing physical models in words is like describing great music in words. Just as you have to listen to music to appreciate it, you have to read the equations of physics to grasp their meaning. This has resulted in a sharp division between "knows" and "know-nots," drawn at such a very high level of knowledge that it excludes the great bulk of humanity—from the person on the street to otherwise highly educated academics who have neglected their mathematical studies. Even many PhD physicists and astronomers whose specialties are in fields other than particle physics and cosmology do not fully appreciate these new developments because they are cast in a language that only the experts speak.

In the main text of this book, I lay out the arguments without mathematical details so that the general reader can at least get a sense of the gigantic conceptual developments that took place over the last century. In mathematical supplements to the main text, I attempt to lower the dividing line in mathematical sophistication needed for a deeper understanding, so that a significantly greater number of people (though, unfortunately, not everyone) can more clearly view the portrait that now exists of the fundamental nature of the Universe.

The mathematics used in the supplements are of the current undergraduate university curriculum in the United States. In most cases, I only utilize the algebra and calculus of the first two years in the typical American university. However, in some places the more advanced methods of the second two years are used. Certainly, anyone who has taken the major courses in a four-year curriculum of undergraduate physics, chemistry, engineering, or mathematics should have no trouble with the material presented, especially if this has been supplemented by reading of popular and semipopular books on modern physics and cosmology. For the less mathematically sophisticated, I present all the arguments in words in the main text; however, by at least glancing at the equations, you will perhaps gain some further insight into what has been one of the greatest achievements in human history.

I have not utilized any quantum field theory. I highly recommend the textbook by F. Mandl and G. Shaw[6] for the student who wishes to move on to the next level. While the mathematics of quantum field theory does not extend much beyond what I have used in the mathematical supplements here, the algebraic calculations are very lengthy and time-consuming to work out in detail, and generally demand the discipline of a graduate course.

A popular book that can provide further background for the material covered here is *Symmetry and the Beautiful Universe* by Leon M. Lederman and Christopher T. Hill.[8] Other references are sprinkled throughout the text.

I am very grateful to the members of the Internet discussion group avoid–L@hawaii.edu, who provide me with continual feedback on my work. I would specifically like to thank Richard Carrier, Marcus Chown, Lawrence Crowell, Keith Douglas, Pedro Hernandez, Kristjan Kannike, Bruce Kellett, Norman Levitt, Donald McGee, Lester Stacey, John Stone, Edward Weinmann, William Westmiller, and especially Taner Edis, Yonathan Fishman, Ludwig Krippahl, Brent Meeker, and Bob Zannelli, for their thoughtful, detailed, and invaluable comments. I greatly appreciate the support I have received over the years from Paul Kurtz, Steven L. Mitchell, and the staff of Prometheus Books. Finally, I would accomplish nothing without the loving support of my wife, Phylliss, our children Noelle and Andy, and their young families.

NOTES

1. Thomas Kuhn, *The Structure of Scientific Revolutions* (Chicago: University of Chicago Press, 1970).

2. Steven Weinberg, "The Revolution That Didn't Happen," *New York Review of Books*, October 1998.

3. Kip S. Thorne, *Black Holes & Time Warps: Einstein's Outrageous Legacy* (New York: Norton, 1994), p. 86.

4. Bede Rundle, *Why There Is Something rather than Nothing* (Oxford: Clarendon, 2004).

5. Ibid.

6. Victor Stenger, *The Unconscious Quantum: Metaphysics in Modern Physics and Cosmology* (Amherst, NY: Prometheus Books, 1995).

7. F. Mandl and G. Shaw, *Quantum Field Theory*, rev. ed. (New York: Wiley, 1984).

8. Leon M. Lederman and Christopher T. Hill, *Symmetry and the Beautiful Universe* (Amherst, NY: Prometheus Books, 2004).

CHAPTER ONE

WHAT ARE THE LAWS OF PHYSICS?

A PRAGMATIC PROCESS

Philosophers of science have never achieved a consensus on what constitutes science. Still, most scientists have a good idea, even if they cannot lay it out in precise terms. Fundamentally, science deals with observations and their descriptions in terms of *models*. A model is more than simply a photograph-like image of a specific set of data. It should be able to successfully describe in a repeatable, testable fashion a whole class of observations of the same general type; enable the prediction of other unexpected observations; and provide a framework for further applications, such as in technology or medicine.

Well-developed and well-tested models of especially universal value are often called *theories*. However, as I indicated in the preface, I will try to avoid this term since it conjures up different and often conflicting meanings in people's minds. While some scientists and philosophers regard well-established scientific theories as the ultimate Platonic reality of the Universe, some laypeople think of them as little more than idle speculations. At least the term *model* avoids both extremes and unnecessary philosophical disputation.

Physical models usually constitute a set of mathematical and logical procedures called *algorithms*. The driving force behind their development is not unquestioned authority, impeccable mathematical logic, or some inexorable drive toward progress. Rather, the models of physics and the other physical sciences evolve by a pragmatic process that exhibits many Darwinian elements.

First, physical models must work. They must agree with the data, pass stringent tests, and be capable of yielding useful results, as described above. Those that do not are rejected; they fail to survive. And since scientific instrumentation becomes increasingly precise as technology advances, models must often undergo revision in order to keep pace.

Second, models should be as simple as possible. They must be parsimonious in their hypotheses. When alternate models are available, those that make the fewest assumptions survive. But, they still must agree with the data. As Einstein said, "Things should be made as simple as possible, but not simpler."

Third, models must be novel. They should tell us something we do not already know and do something new for us we cannot already do. Scientists pay little attention to many new models that are proposed by outsiders, not because they come from outside the club, but because they inevitably tell us nothing new, make claims that cannot be tested, or have already been proven wrong. For example, hundreds if not thousands of books and articles have been written claiming that "Einstein was wrong." Yet Einstein's relativity, which is a century old at this writing, remains intact because no confirmed, repeatable experiment or observation has yet proven it wrong.[1]

The process of scientific development does not ensure that the models agreed upon are unique or the best anyone can come up with. And the process certainly does not mandate that any currently accepted model will forever remain immune from being rejected due to its failure to agree with some future observation. Often new models are proposed that improve on older ones by simplifying the assumptions or placing them on a firmer foundation. But *empirical falsification* is the primary measure by which previously successful models are discarded. Although falsification is neither necessary nor sufficient for distinguishing science from nonscience, as is sometimes claimed, it remains a crucial element of the scientific process.

None of this should be taken to mean that any one model is as good as another. Neither does it imply that models are simply speculations. As already indicated, a scientific model must agree with the data and pass stringent empirical tests before being accepted by the scientific community. Relativity and quantum mechanics are prime example of models that have been successfully tested many times with great precision. While single experiments have occasionally been reported that claimed to falsify either model, none have been confirmed by further investigations.

In this chapter, I will briefly review the basic principles of physics contained in our models, as they are currently formulated. While I will more or less follow the historical sequence, I will not provide a detailed discussion of how these developments came about. I hope the reader will seek out that history from the many other sources that are available. My purpose at this point is to provide a concise description that can be referred back to in the following chapters.

CLASSICAL PHYSICS

Let me begin by summarizing what are currently recognized as the most important laws of classical physics, where by classical physics I mean physics prior to the twentieth century. Foremost are the great conservation principles for energy, linear momentum, and angular momentum. I assume that the reader is familiar with these and the other basic quantities of physics, such as velocity and acceleration, which are defined in the most elementary physics textbooks.

Conservation of energy: When a physical system is sufficiently isolated from its environment, the total energy of the bodies in that system is a constant. The bodies within the system can interact and exchange energy, but the total energy remains fixed.

Conservation of linear momentum: When a physical system is sufficiently isolated from its environment, the total linear momentum of the bodies in that system is a constant. The bodies within the system can interact and exchange linear momentum, but the total linear momentum remains fixed.

Conservation of angular momentum: When a physical system is sufficiently isolated from its environment, the total angular momentum of the bodies in that system is a constant. The bodies within the system can interact and exchange angular momentum, but the total angular momentum remains fixed.

Now, not every physical system is observed to obey these conservation principles. When a nonisolated system does not exhibit energy conservation, we describe that empirical fact as resulting from either the flow of heat energy into (or out of) the system, or work done by (or on) the system, or both. In this form, conservation of energy is known as *the first law of thermodynamics*.[2] According to the first law, the change in the internal energy of the system—the energy it would have if it were isolated—is equal to the heat input minus the work done by the system. That change can be an increase or a decrease, depending on the algebraic signs of the heat and work; input heat is positive, output heat is negative, work done by the system is positive, and work done on a system is negative.

When a physical system does not exhibit total linear momentum conservation, we describe that empirical fact as a net force acting on the system, where the net force is defined as the time rate of change of the total linear momentum. This is *Newton's second law of motion* (Isaac Newton, d. 1727). When the mass of a body is constant, Newton's second law says that the net force on a body equals the product of the body's mass and acceleration, as given by the famous formula $\mathbf{F} = m\mathbf{a}$. We can view mass as the measure of a body's static inertia, while momentum is a measure of its dynamical inertia or a more general "quantity of motion" that takes into account both mass and velocity.

Newton's first law of motion says that a body at rest will remain at rest and a body in motion at constant velocity will remain in motion at constant velocity unless acted on by an external force.

Newton's third law says that when two bodies interact, the force of body 1 on body 2 is equal and opposite to the force of body 2 on body 1. Both of these laws follow from conservation of linear momentum.

When a physical system does not exhibit total angular momentum conservation, we describe that empirical fact as a net *torque* acting on the system, where the net torque is defined as the time rate of change of the total angular momentum. Laws of rotational motion similar to Newton's laws of rectilinear motion, described above, can then be derived.

Newton's laws implement a principle that had been elucidated by Galileo (Galileo Galilei, d. 1642) a generation earlier. When Galileo, following Copernicus (Nicolaus Copernicus, d. 1543), convinced himself that Earth moved around the Sun, he had to explain why we do not notice this motion. He realized that we can sense only changes in velocity, not absolute velocity itself. That is, velocity is relative. Let me state this principle in the following way.

Principle of Galilean relativity: There is no observation that enables you to distinguish between being in motion at constant velocity and being at rest.

Since the laws of physics must describe observations, another statement of the principle of Galilean relativity is that the laws of physics must be the same in all reference frames moving at constant velocity with respect to one another.

Aristotle (d. 322 BCE) had taught that a force is necessary for motion, as it seems in commonsense experience. Galileo showed this was wrong, and Newton's second law expresses the fact that a force is not needed to produce a velocity, but rather to produce a change in velocity, or, more precisely, a change in momentum.

The mathematical methods of classical mechanics provide a way to predict the motion of a body. By knowing the initial position and momentum of a body, and the forces acting on it, we can predict the future position and momentum of that body. In order to know the forces, however, we need some laws of force. Newton also provided the law of force for gravity.

Newton's law of gravity: Two particles will attract one another with a force proportional to the product of their masses and inversely proportional to the square of the distance between them.

Using the mathematical technique *calculus*, which Newton and Gottfried Leibniz (d. 1716) invented independently, Newton was able to show how to calculate the force between two bodies of arbitrary shape by adding up the particulate contributions. He proved that spherical bodies, such as Earth and the moon, attract one another as if their masses were concentrated at their centers, a great simplification for bodies, like human beings, that are found near the surface of a large planet. Of course, most humans are not spheres, but can be treated as pointlike on the planetary scale.

Prior to the twentieth century, the only known fundamental forces were gravity, electricity, and magnetism. The force law for static electricity is similar to the law of gravity, with the added twist that it can be repulsive as well as attractive. Gravity is also repulsive, as we will see in a later chapter.

Coulomb's law: Two electrically charged particles will attract or repel one another with a force proportional to the product of their charges and inversely proportional to the square of the distance between them (Charles-Augustin de Coulomb, d. 1806).

Attraction occurs when the charges have unlike algebraic signs; repulsion occurs when the charges have like algebraic signs. Again, calculus can be used to compute the force between two extended charged bodies.

The static magnetic force depends on electric current, the flow of electric charge. In its simplest form, we consider two long, straight, parallel wires.

Magnetic force: The magnetic force per unit length between two long, straight, parallel wires is proportional to the product of the currents in the wires and inversely proportional to the distance between them. The force is attractive when the currents flow in the same direction and repulsive when they flow in opposite directions.

The magnetic force in more complicated situations can be calculated from a more general law using vector calculus that need not be detailed here.

At this point, we must add another conservation principle.

Conservation of electric charge: The total electric charge of isolated systems is constant.

A familiar application of conservation of charge is in electric circuits, where the total current into any junction must equal the total current out. Another well-known example is in chemical reactions. For example, $H^+ + H^+ + O^{--} \rightarrow H_2O$, where the superscripts indicate the net charge, positive or negative, of hydrogen and oxygen ions, the resulting water molecule is electrically neutral.

In describing gravitational, electric, and magnetic forces, physicists utilize the notion of a *field*. A field is a mathematical quantity that has a value at every point in space. If that quantity is a single number, it is called a *scalar field*. If the field also has a direction, it is called a *vector field*. The classical gravitational, electric, and magnetic fields are vector fields. Even more numbers are needed to specify *tensor fields*. The gravitational field in Einstein's general relativity is a tensor field.

Fields are often pictured as a series of "lines of force" that surround a body that is regarded as the source of the field. In a common classroom demonstration, iron filings are sprinkled on a piece of paper that is placed over a bar magnet. The fillings line up along the so-called magnetic field of the magnet. However, this experiment should not be interpreted as demonstrating the existence of a "real" magnetic field. Electrodynamic calculations can be done without ever mentioning fields; they are simply useful constructs.

Experiments in the eighteenth century showed that electricity and magnetism were closely interrelated, that one could produce the other even in the absence of charges and currents. A set of equations developed by James Clerk Maxwell (d. 1879) allowed for the calculation of the electric and magnetic fields for any charge and current configuration. Even more profoundly, the equations suggested a model of *electromagnetic waves* in empty space that travel at the speed of light. This indicated that light was itself an electromagnetic wave and predicted that waves of frequencies outside the visible spectrum should be observed, produced by oscillating charges and currents. Shortly thereafter, radio waves were produced in the laboratory and found to move at the speed of light, marvelously confirming that prediction.

Although the new electromagnetism was spectacularly successful, a question remained: What is the medium that is doing the waving? Sound waves were known to be vibrations of a material medium, such as air, that propagated through the medium; the wave equation for electromagnetism was of the same form as that for sound waves. Consequently, it was widely assumed that an invisible, elastic medium called *ether* pervaded all of space. Electromagnetic waves were, then, vibrations of this ether. But this picture was found wanting by developments in the early twentieth century.

Before we get to that, another item needs be added to complete the story of the fundamental laws physics prior to the twentieth century.

The second law of thermodynamics: The total entropy of an isolated system can only stay the same or increase with time.

This law expresses the empirical fact that many processes appear to be irreversible. Air rushes out of a punctured tire, never in. Broken glasses do not spontaneously reassemble. Dead people do not come alive.

Entropy is vaguely associated with disorder, although it has a perfectly precise mathematic definition. The second law of thermodynamics expresses the commonsense intuition that systems left to themselves become more disorderly with time, eventually reaching equilibrium with their surroundings. Of course, we know that a system can become more orderly with time, such as when we build a house. But this requires the expenditure of energy. When two systems are connected, the entropy of one can decrease, provided the entropy of the other increases by the same amount or more. While the entropy of a house decreases during its construction, everything around the house experiences an entropy increase that must be hauled away to the garbage dump.

The reader may be surprised to hear that the second law of thermodynamics is not a universal law of physics. It does not usefully apply, for example, at the molecular level. The second law represents a statement about the average behavior of systems of many particles, specifically their tendency to approach equilibrium. However, at the molecular level and below (the atomic, nuclear, and subnuclear levels), random fluctuations can result in a spontaneous decrease in the entropy of an isolated system. These fluctuations are not noticeable for the bodies of normal experience, which typically contain a trillion trillion (10^{24}) molecules.

Furthermore, as shown by Ludwig Boltzmann near the end of the nineteenth century, the second law of thermodynamics is simply a macroscopic convention for what we take to be the direction or "arrow" of time. This is a profound issue that will be elaborated on later in this book.

Summarizing to this point, the conventionally accepted basic principles of classical physics are Newton's laws of motion; the principle of Galilean relativity; the gravitational, electric, and magnetic force laws; Maxwell's equations of electrodynamics; and the principles of conservation of energy, linear momentum, angular momentum, and electric charge. Also included are the first and second laws of thermodynamics, where the first law is another form of conservation of energy and the second law is simply a definition of the conventional direction of time. These principles play a basic role across the board in physical phenomena. Much more classical physics exists, such as fluid mechanics, wave mechanics, and thermodynamics. However, their principles can be derived from the fundamental principles I have summarized. And, most important, these laws and principles successfully described all the scientific observations of material systems that had been made up until the twentieth century, with just a few exceptions that I will discuss later in this chapter.

THE MULTIPLE METHODS OF MECHANICS

During the two centuries following the publication of Newton's *Principia*, classical mechanics evolved to a high state of perfection. Mathematical methods were developed by which the equations of motion could be derived for any system of bodies with any number of degrees of freedom. In these methods, which are still in wide use today, generalized spatial coordinates are defined that could be the familiar Cartesian coordinates or other types of spatial coordinates, such as angles of rotation about various axes.

In the method developed by Joseph Louis Comte de Lagrange (d. 1813), a *Lagrangian* is defined (in most familiar cases) as the difference in the total kinetic and potential energies of the system, which depend on the generalized coordinates and velocities. The equations of motion are then determined by a set of differential equations called Lagrange's equations.

Another generalized method for determining the equations of motion was developed by William Rowan Hamilton (d. 1865). We define a *Hamiltonian* as, in most applications, simply the total energy of the system; that is, the Hamiltonian equals the sum of the kinetic and potential energies. These are taken to be functions of the generalized coordinates and the so-called *canonical momenta* corresponding to each coordinate. Then, another set of differential equations called Hamilton's equations of motion can be used to give the law of motion for that system.

In yet another generalized method, also introduced by Hamilton, we define a quantity called the *action* as a product of the Lagrangian averaged over the time that it takes the body to move over some path from a point *A* to a point *B*, multiplied by that time interval. The path predicted by the equations of motion is precisely the path for which the action is *extremal*, that is, either maximum or minimum. Usually this is called the *principle of least action* since it expresses the intuition that a body takes the "easiest" path in going from one point to the next. Least action is also called *Hamilton's principle*, but it should not be confused with the Hamiltonian method using his equations of motion. In the twentieth century, Richard Feynman generalized the least action principle to produce a version of quantum mechanics called *path integrals* that has many advantages over more conventional methods.[3]

THE CLOCKWORK UNIVERSE

Classical mechanics enables us to predict the future motion of a body from knowledge of its initial position, initial momentum, and the net force acting on the body. The usual philosophical interpretation given to this procedure holds that a causal process is taking place, where the force is the causal agent. The final position and momentum at some later time is then fully determined by that force and the initial conditions. Since the Universe, as far as we can tell, is composed solely of localized bodies, the initial positions and velocities of those bodies and the forces acting on them predetermine everything that happens. This notion is referred to as the *Newtonian world machine* or the *clockwork universe.*

Now, as already mentioned, the time-reversed situation is also perfectly compatible with Newtonian mechanics and, indeed, all of physics as we know it today. Thus, we can take a final position and momentum of a body, and the net force, and "retrodict" its initial quantities. This is the process astronomers use to take observations made today to determine where a particular star may have been in the sky back in, say, 2000 BCE. But what does it mean? Is the Universe determined by what happened in the past, or is it determined by what is to happen in the future?

The existence of these two alternate procedures does not cause most physicists or astronomers to lose any sleep. However, it presents a serious problem for anyone trying to extract some metaphysical interpretation—especially when one is wedded to an ontology based on cause-and-effect and one-way time. Contrary to commonsense, the laws of mechanics imply that time can run either way and that what we label as the cause and what we label as the effect are interchangeable when describing the motion of a single particle. As mentioned above, the familiar arrow of time is a statistical effect that applies only for systems of many particles with considerable random motions, such as most of the physical systems of normal experience. Time-directed cause-and-effect may be an appropriate concept in that domain, but should not be assumed to apply at the fundamental level of quantum phenomena. This does not mean that no type of causal connection applies, just that this connection must not distinguish between cause and effect.

The interpretational problems of classical mechanics become even more serious when we consider the least-action approach. There, both the initial and the final spatial positions are assumed and the precise path followed from one to the other is determined. If you require your interpretation to involve a time-directed casual process, the particle, or the force acting on it, has to somehow "know" where the particle is going to wind up so that it travels the correct path.

As long as each method reproduces measurements accurately, none can be said to be wrong. Indeed, a given method can be mathematically derived from any of the others. If that is the case, then which method is the most fundamental? Which describes "reality" as it actually exists and which is just a mathematical transformation of that "true" description? When we try to extract some kind of philosophical truth from physics, we need to take into account the fact that different but equally successful models may imply different and possibly incompatible metaphysical views.

SPECIAL RELATIVITY

By the end of the nineteenth century, classical physics had successfully described most of what was observed in the physical world until that time. However, there were a few mysterious anomalies. Explaining these would entail two major revisions in physics thinking.

One anomaly was a theoretical one. Maxwell's equations implied that electromagnetic waves in a vacuum moved at the speed of light, with no account taken of the motion of the source or observer. Indeed, if electromagnetic waves corresponded to vibrations of a material ether, that ether formed an absolute reference frame for the Universe with respect to which all bodies moved. This contradicted the principle of Galilean relativity, which says that no such reference frame exists.

In the 1890s, Albert Michelson and Edward Morley attempted to measure the speed of Earth through the ether. That speed should have changed by 60 kilometers per second as Earth turned around the Sun in the course of half a year. The experiment was fully capable of detecting that motion to a hundredth of a percent. However, no change in speed was observed.

In 1905 Albert Einstein examined the consequences of assuming that the speed of light stays constant while relativity remains valid. He found this required a change in our common views of space and time, and a rewrite of some of the basic equations for physical quantities, such as momentum and kinetic energy.[4]

Einstein found that the distance interval between two points in space and the time interval between two events at those points are different for observers moving with respect to one another. That is, if I were to measure distance and time intervals with a meterstick and clock sitting here on Earth, and you were to measure them with your own meterstick and clock inside a spaceship going at a very high speed relative to Earth, we would arrive at different values. I might see two supernova explosions simultaneously, while you might see one occurring before the other. We do not notice this disagreement in everyday life because we experience relatively low speeds compared to the speed of light, where the effect is unnoticeable.

GENERAL RELATIVITY

Einstein's *general relativity* is basically a model of gravity. A certain tiny anomaly in the motion of the planet Mercury observed by astronomers in the nineteenth century suggested that Newton's law of gravity was not fully adequate. In 1915 Einstein was able to show that the acceleration of a body under the action of gravity can be described as the natural motion of the body in a curved, non-Euclidean space. In addition to successfully calculating the Mercury anomaly, Einstein predicted that light would be bent by gravity, for example, as it passes by the Sun. He also proved that a clock in a gravitational field will appear from the outside to run slower. This, in turn, implied that electromagnetic waves such as light passing through the field decrease in frequency, that is, appear "redshifted." These phenomena have all been confirmed. After almost a century of stringent tests, general relativity remains consistent with all observations.[5]

QUANTUM MECHANICS

As the nineteenth century ended, the wave model of light had other problems besides the apparent absence of evidence for any medium to do the waving. The wave model incorrectly described the wavelength spectrum of light radiated from heated bodies. It also failed to account for the photoelectric effect in which light induces an electric current.

Over about a twenty-year period in the twentieth century, a drastically different model called *quantum mechanics* was developed to describe these and a host of other novel phenomena that were being observed as laboratory experiments became increasingly sophisticated.

As suggested by Max Planck in 1900, light was found to occur in discrete bits of energy called *quanta*. In 1905 Einstein identified these quanta with individual particles we now call *photons*. In 1923 Louis de Broglie suggested that all particles, such as electrons, exhibited the same kind of wavelike behavior that we see with light and this was shortly confirmed in the laboratory. This observation of wavelike behavior in particles has come to be known as the *wave-particle duality*. Physical bodies seem to behave as both particles and waves. Quantum mechanics gradually developed as a mathematical procedure for handling these so-called particle-waves, or "wavicles." Even today, much mystery is attributed to the wave-particle duality. However, modern physicists have been able to reconcile the two views mathematically and much of the mystery is in the eyes of the beholder who is looking for mystery.

Today quantum mechanics exists as a very successful model, but the axioms from which the model is deduced largely deal with its mathematical formalism and it is difficult to say what new physical laws are implied independent of those structural assumptions.

If we try to single out the most significant new concept introduced in quantum mechanics, it would probably be the realization that to observe an object we must necessarily interact with that object. When we observe something, we must bounce photons (or other particles) off it. Photons carry momentum, and so the object being observed recoils in the process. While this recoil is negligible for most common, everyday observations, such as looking at the moon, it becomes very important at the atomic and subatomic levels, where "looking" at an atom can split the atom apart. When we try to look at an atom with increasing spatial resolution, the momentum of that atom becomes increasingly uncertain as we impart greater momentum to that atom. This is codified in the following famous principle.

The Heisenberg uncertainty principle: The more accurately we try to measure the position of an object, the less accurately can we measure the momentum of that object and vice versa.

The uncertainty principle has profound consequences for the following reason: As we saw above, classical mechanics provides a procedure for predicting the motion of a body. Given the initial position and momentum of a body and knowing the forces on that body, we can predict the future position and momentum of that body. That is, we can calculate the trajectory of a body to any degree of accuracy. The philosophical implication of classical mechanics, then, is that the motions of all the bodies in the Universe are determined by prior events (or, perhaps, future events). The Universe is thus just a giant machine—the Newtonian world machine. Or, in another commonly used metaphor, we live in a clockwork universe.

The Heisenberg uncertainty principle implies that, in fact, we do *not* live in a machinelike, clockwork universe. The motions of bodies are not all predetermined, which means that they have some random element in their motion. Quantum mechanics implements this conclusion by providing procedures for computing the *probabilities* for a physical system to move from some initial state to some final state. In fact, the classical laws of motion remain unchanged, as long as you apply them to the average values of ensembles of measured quantities and not the individual measurements themselves.

Quantum mechanics has done much more than simply replace classical mechanics with corresponding statistical calculations. New phenomena, such as the *spin* or intrinsic angular momentum of point particles, arise only in a quantum context.

For completeness, I must mention an alternate interpretation of quantum mechanics, proposed in the 1950s by David Bohm, in which the apparent indeterminism of quantum phenomena results from subquantum forces.[6] No evidence for such forces has yet been found, and Bohm's model violates special relativity in requiring superluminal processes. These facts make Bohm's model a generally unpopular proposal among physicists, although it still retains some measure of support.

I discuss the various philosophical interpretations of quantum mechanics in my books *The Unconscious Quantum*[7] and *Timeless Reality*,[8] which can be consulted for further references.

THE STANDARD MODEL

Attempts in the 1920s and 1930s to combine special relativity and quantum mechanics met with some success, especially with the work of Paul Dirac, which provided a basis for spin and predicted the existence of antimatter. However, a full uniting of the two models was not achieved until after World War II with the development of *quantum electrodynamics*. This proved enormously successful and produced exquisite calculations that agree with equally exquisite measurements to many significant figures.

Quantum electrodynamics, or QED, is part of a more general mathematical structure called *relativistic quantum field theory*.[9] QED deals only with the electromagnetic force. Continuing attempts to bring gravity into the quantum regime, uniting general relativity and quantum mechanics, have proved highly elusive. Even today, no such unified theory exists, although physicists are pursuing several lines of attack, such as *string theory*, in a relentless quest to achieve the goal of unification.[10] The main obstacle has been the lack of any guidance from experiment. Even with the highly sophisticated instruments of current technology, we are still far from being able to empirically explore the effects of quantum mechanics on gravity. In the meantime, we have two additional forces to contemplate.

Early in the twentieth century, experiments had elucidated the structure of atoms and found that each was composed of a very compact nucleus of protons and neutrons, surrounded by a nebulous cloud of electrons. Apparently, a previously unknown force called the *strong force* was holding the "nucleons" in the nucleus together. It was also observed that some nuclei spontaneously emitted electrons in a process known as *beta-decay* (β-decay). A fourth force, the *weak force*, was found to be responsible for this type of radioactivity. We now know that the Sun and, consequently, life on Earth are powered by the weak force.

Furthermore, as particle accelerators became increasingly powerful, many new particles were discovered. In the 1970s we learned that the protons and neutrons inside nuclei were composed of more fundamental objects called *quarks*. The electron remained fundamental, but was found to be one in another class of particles called *leptons*, which included two other heavier, electron-like particles, the *muon* and *tauon*, and three electrically neutral particles of very low mass, the *neutrinos.* All of these are accompanied by their antiparticles, which are particles of the same mass and opposite charge.

After an immense worldwide effort involving thousands of experimentalists and theorists, a new model was developed called the *standard model.*[11] In this model, the universe is composed of quarks and leptons interacting with one another by the electromagnetic, strong, and weak nuclear forces. Since it is so much weaker, comparatively, than these forces, gravity can be largely neglected at the subnuclear scale, at least at current levels of experimentation.

The forces in the standard model are visualized as resulting from the exchanging back and forth among quarks and leptons of another set of "mediating" particles called *gauge bosons.* The photon is the gauge boson mediating the electromagnetic force. The strong force is mediated by *gluons.* The weak force is mediated by the exchange of W- and Z-bosons. Furthermore, the electromagnetic force and weak force are united in a single force called the *electroweak* force.

Those disciplines within physics dealing with collective phenomena have produced their own parallel advances of immense import: solid-state electronics, superconductivity, and Bose condensation, to mention a few. While not inconsistent with the standard model, the associated phenomena are not derived directly from it. Similarly, most of the details of chemistry, biology, and other natural sciences are described by their own principles. These details are understood to have "emerged" from the interactions of fundamental particles, but with a large element of chance that makes their precise forms and structures far from inevitable or predictable. As a result, experts in these fields do not need to know much particle physics, even though particles underlie the phenomena they study.

Although, at this writing, the standard model remains in agreement with all observations, no one believes it to be the final word. Depending on how you count, twenty-four or so standard model parameters are undetermined by the model and must be measured in experiments. These include the masses of almost all the fundamental particles and the relative strength of the various forces, which most physicists believe will ultimately be united as one. While this may seem like a large number, remember that hundreds of precise experimental results can be fit to the standard model.

The standard model more or less reached its current form in the 1970s. Since then, theoretical physicists have been busily trying to unearth a more basic model from which the current model can be derived. This quest has so far been unsuccessful, although several leads, especially string theory, are being vigorously pursued. As the history of physics has shown, most progress is made when experiment and theory (model building) work hand in hand, each suggesting to the other where to look for new data and new ideas. So far, experimental physicists have been unable to provide sufficient empirical data needed to guide theoretical developments beyond the standard model. The cost of pushing back the frontier of fundamental physics has become almost prohibitively expensive, but good reasons exist to push on nevertheless. However, I would like to add a note of caution. Even if all the needed funding became available, it could turn out that the standard model is no more derivable from a deeper model than the rest of science is derivable from the standard model. Perhaps many of the details of the standard model also emerged by a random process that might have taken many different paths. This is a possibility that is explored in this book.

NOTES

1. Clifford M. Will, *Was Einstein Right? Putting General Relativity to the Test* (New York: Basic Books, 1986).

2. Some physicists assert that the first law of thermodynamics applies only to macroscopic systems, but I see no reason for this artificial distinction. However, the second law of thermodynamics is generally meaningful only for large numbers of bodies, as in the case macroscopically.

3. Richard Feynman, *The Principle of Least Action in Quantum Mechanics* (Ann Arbor, MI: University Microfilms, 1942).

4. Many biographies of Einstein exist. See, for example, Abraham Pais, *"Subtle is the Lord . . .": The Science and the Life of Albert Einstein* (Oxford: Oxford University Press, 1982).

5. Will, *Was Einstein Right?*

6. D. Bohm and B. J. Hiley, *The Undivided Universe: An Ontological Interpretation of Quantum Physics* (London: Routledge, 1993).

7. Victor Stenger, *The Unconscious Quantum: Metaphysics in Modern Physics and Cosmology* (Amherst, NY: Prometheus Books, 1995).

8. Victor Stenger, *Timeless Reality: Symmetry, Simplicity, and Multiple Universes* (Amherst, NY: Prometheus Books, 2000).

9. For a low-level discussion of QED, see Richard Feynman, *QED: The Strange Theory of Light and Matter* (Princeton, NJ: Princeton University Press, 1985). For a history of QED, see S. S. Schweber, *QED and the Men Who Made It: Dyson, Feynman, Schwinger, and Tomonaga* (Princeton, NJ: Princeton University Press, 1994).

10. Brian Greene, *The Elegant Universe: Superstrings, Hidden Dimensions, and the Quest for the Ultimate Theory* (New York: Norton, 1999).

11. Lillian Hoddeson, et al., eds., *The Rise of the Standard Model: Particle Physics in the 1960s and 1970s* (Cambridge: Cambridge University Press, 1997); Gordon Kane, *The Particle Garden: Our Universe as Understood by Particle Physicists.* (New York: Addison-Wesley, 1995).

CHAPTER TWO

THE STUFF THAT KICKS BACK

"I REFUTE IT THUS"

Most scientists will tell you that they only deal with what is observable in the world. If an unobservable reality underlies their measurements, we have no way of knowing about it.

Still, scientists believe that their observations and the models they build to describe those observations have something to do with an ultimate reality. They reject the notion that scientific models are merely the subjective narratives of their particular culture, as some postmodern authors have asserted. The great success of science and its dominant role in human life belies the postmodern claim.

Besides, each of us has an intuitive feeling that the objects we confront during our waking experience constitute some aspect of reality. In 1793 Samuel Johnson expressed this common view. As described in Boswell's *Life of Johnson*,

> We stood talking for some time together of Bishop Berkeley's ingenious sophistry to prove the nonexistence of matter, and that every thing in the universe is merely ideal. I observed, that though we are satisfied his doctrine is not true, it is impossible to refute it. I shall never forget the alacrity with which Johnson answered, striking his foot with mighty force against a large stone, till he rebounded from it, "I refute it thus."[1]

When we kick an object and it kicks back, we are pretty certain that we have interacted with some aspect of a world outside our heads (and feet). In simple terms, this describes the processes of everyday observations as well as the most sophisticated scientific experiments. When we look at an object with our naked eyes, light from some source bounces off the object into our eyes. Or the object itself may emit light. In either case, the object and the light receptors in our eyes recoil from the momentum that is transferred in the process and generates an electrical signal that is analyzed by our brains.

Scientific observations are basically of the same nature. Not just visible light but the entire electromagnetic spectrum from radio waves to gamma rays (*γ-rays*) is used to joggle reality, along with sensors far more precise than the human eye to detect the jiggles that are returned. What's more, other particles, such as electrons and neutrinos, are also available as probes and computers are utilized to supplement the analytic capability of our brains.

In short, science is not some unique method of learning about the world. It is an enhancement of the *only* method by which we humans, barring divine revelation, learn about the world—empirical observation. In the model with which we describe these observations, the stuff that kicks back when you kick it is called *matter*. We will further assume that matter comes in empirically distinguishable units we call *bodies*. The property of a body that enables it to kick back we call *inertia*. The mass and momentum of a body are measures of its inertia.

THE OBSERVABLES OF PHYSICS

All the quantities physicists call "observables" are operationally defined; that is, they are specified in terms of measurements performed by well-prescribed measuring procedures. Time is what we measure on a clock. Temperature is what we measure with a thermometer. We will return frequently to the role of operational definitions in physics. The main point to keep in mind at this juncture is that these definitions, along with the other mathematical quantities that appear in physics equations, are human inventions. They are parameters used in scientific descriptions of observed events. Those descriptions, or models, are used to reproduce events empirically and predict future events. They provide a framework for our measurements and must be formulated to some degree before we even begin to make those measurements. However, while the components of physical models must be consistent with empirical data, they need not correspond exactly to the underlying elements of reality. This includes—most notably—space and time.

Space and time are commonly considered part of the inherent structure of the Universe. In the view I am adopting here, this is not necessarily the case. Or, at least, I do not require it to be the case and, thus, I do not have to spend a lot of time trying to justify such a claim. Just as a picture of a chair is not a chair, space and time are the paper on which we draw a physicist's picture of reality. We have no way of judging whether they are part of reality itself. Indeed, the philosophical problems associated with developing a quantum theory of gravity, which are beyond the scope of this book, throw into doubt common notions of space and time.[2]

Furthermore, our models are not unique. If we have two (or 10^{100}) models that do equally well in describing the data, which is the "true reality"? While it is difficult to imagine a description of reality that does not include space and time, such a description is not out of the question and some attempts have been made to develop models along those lines. For now, however, let us stick to the familiar notions of space and time, which are deeply embedded in both our scientific culture and everyday lives.

Let us begin our quantitative description of observations with the accepted operational definition of time. Following Einstein, we define time simply as what is measured on a clock. Whatever its construction, a clock will provide a series of ticks and a counter of those ticks. Note that in our operational view, time is fundamentally discrete.

Almost any series of repetitive events can be used as a clock, such as my own heartbeats. However, defining time by my heartbeats would result in physics equations that would have to be written in such a way as to take into account my daily activity. The time it takes an object to fall from a given height would be less on a day that I sat around at the computer than on a day when I played three sets of tennis. Obviously, for the sake of simplicity and the desire to describe reality in an objective way, we should use some more universal measure of time.

Throughout history, time has been measured by highly repeatable, nonsubjective events such as day and night, the phases of the moon, and other observations in the heavens. The Egyptians needed a way to predict the rise of the Nile. They found that counting the moon's cycles was inadequate and around 4326 BCE discovered that the Dog Star in Canis Major (now called Sirius) rose near the Sun about every 365 days about the time of the annual Nile flood.

Quite a bit later, but still in what we view as ancient times, in 46 BCE Julius Caesar introduced the Julian calendar, which included a leap day every four years. This was corrected further by eleven minutes per year in 1582 with the Gregorian calendar, which is still in use today.

While astronomical calendars, divided into days and hours, suffice for the timing of familiar events, science has required a finer scale. Galileo introduced the pendulum as a timing device for experiments, and even today mechanical clocks and watches rely on the pendulum to provide uniformity and accuracy. However, science has greatly improved the methods for standardizing time and the mechanical clock has become obsolete, except as an attractive piece of antique jewelry or furniture.

Our basic unit of time, the second, was historically defined as 1/86,400 of a mean solar day. In 1956 this was changed by international agreement to "1/31,556,925.9747 of the tropical year for 1900 January 0 at 12 hours ephemeris time." However, defining time in terms of astronomical observations eventually proved inadequate. A clock was needed that could, in principle, sit by itself in a laboratory and not have to be constantly recalibrated against Earth's motion around the Sun.

In 1967, again by international agreement, the second was redefined as the duration of 9,192,631,770 periods of the radiation corresponding to the transition between the two hyperfine energy levels of the ground state of the Cs^{133} atom. In 1997 this was further refined to specify that the second so defined referred to a cesium atom at rest at absolute zero. Since observing such an atom is impossible, this requires a theoretical extrapolation to get the above value. Is this the final standard? Probably not.

Once we have a measure of time, we next need a measure of space or distance. The international standard unit of distance is the *meter*. In 1793 the meter was introduced as 1/10,000,000 of the distance from the pole to the equator. In 1889 the standard meter became the length of a certain platinum-iridium bar stored under carefully controlled conditions in Paris. In 1906 the meter was redefined as 1,000,000/0.64384696 wavelengths in air of the red line of the cadmium spectrum. In 1960 it became 1,650,763.73 wavelengths in a vacuum of the electromagnetic radiation that results from the transition between two specific energy levels (2p10 and 5d5) of the krypton-86 atom. Finally, by international agreement in 1983, the meter was defined to be the distance traveled by light in vacuum during 1/299,792,458 of a second.

This last definition of the meter has a profound consequence that is rarely emphasized in the textbooks. Since 1983, distance in science is no longer treated as a quantity that is independent of time. In fact, distance is now officially defined in terms of time, that is, a measurement fundamentally made not with a meterstick but with a clock: the time it takes light to travel between two points in a vacuum. Of course, in practice we still use metersticks and other means to measure distance, but in principle these must be calibrated against an atomic clock. The National Institute of Standards and Technology (NIST) and the United States Naval Observatory now provide a time source accurate to 100 nanoseconds as a satellite signal.[3] However, the primary time and frequency standard is now provided by a *cesium fountain* atomic clock at the NIST laboratory in Boulder, Colorado, that has an uncertainty of 1×10^{-15} second.[4]

Notice that nowhere in the above discussion is any reference made to some metaphysical notion of time. Although most people have an innate feeling that time is something "real," the requirement of the reality of time is not formally built into the model. Of course, the use of a clock to measure time is based on common experience, and innate feelings are formed from observations. But the beauty of the operational nature of model building is that we can objectively codify observations without reference to "feelings."

The operational view also helps us to come to grips with the fact that models change as science improves its instruments and methods. We do not have to waste time worrying about whether the time measured with Galileo's pendulum is the "same time" as that measured with atomic clocks. We do not have to fret over trying to prove that any particular standard we may adopt is "truly uniformly repetitive." Uniform repetitiveness is defined by our standard. If that standard is my heartbeats, then they are uniformly repetitive *by definition* while the vibrations of cesium depend on my personal activity.

In developing his special theory of relativity in 1905, Einstein postulated that the speed of light in a vacuum is a constant—the same in all reference frames. Since then, special relativity has proved enormously successful in describing observations. As a result, Einstein's postulate is now built into the structure of physics by the very definition of distance.

One often reads stories in the media about some observation or theory suggesting that the speed of light is not a constant. Such claims are meaningless unless distance is first redefined. According to the current standard, the speed of light in a vacuum is a constant—by definition.

However, note that what we call "the speed of light in a vacuum," c, is more precisely the limiting speed of Einstein's relativity. If the photon had nonzero mass, then it would travel at less speed than c. Still, the empirical limit on the photon mass is very small. Also, since the vacuum is not really empty, light might bounce around, resulting in an effective speed less than c, as it does in a medium such as air or glass. Finally, the group velocity of a pulse of light can be less than or even greater than c. None of this changes that fact that c is an arbitrary number.

So, as it now stands by international agreement, all physics measurements reduce to measurements made on clocks. Furthermore, since the process of measuring time involves counting ticks, observables should be represented by rational numbers—integers or ratios of integers. The smallest time interval that can be measured is 5.4×10^{-44} second, the *Planck time*. In Planck units, any value of time is an integer.

Since the Planck time interval is so small, even at the current levels of exploration in the subatomic domain, it becomes possible to approximate time as a continuous variable. Only in quantum gravity and discussions about the very early Universe need we worry about the ultimately discrete nature of time. The same can be said about space, where the minimum measurable distance is 1.6×10^{-35} meter, the *Planck length*.

Being able to approximate distance, time, and the other quantities derived from them as continuous variables, we can take advantage of the mathematical techniques of calculus and its advanced forms such as partial differential equations. Almost all of the laws of physics are expressed in these terms, with the added tools of vector, tensor, and matrix algebra. These mathematical methods are taught at the university undergraduate level and are very familiar to physical scientists and mathematicians.

Besides observables, the equations of physics contain other mathematical objects that are not limited to the set of real numbers. Thus we have complex numbers (numbers that contain $i = \sqrt{-1}$) such as the quantum wave function, abstract space vectors such as the quantum state vector, tensors such as the curvature of space-time, and many others. These are more abstract than measured quantities, though still defined in terms of measurements, and should not be assumed to have any great ontological significance. We will not be using any of these advanced mathematical techniques in the main text, though the supplementary material applies mathematics at about the senior university level or lower.

THE SPACE-TIME MODEL

In the 1930s astronomer Edward Arthur Milne described a procedure by which we can make observations using only a clock to provide quantitative measurements.[5] This procedure implements the notion that objective reality responds to our probing, that is, sending out signals and waiting for replies, and that our familiar space-time picture of reality is basically a model we have invented to describe this process.

Milne argued that as observers of the world around us, all any of us do is transmit and receive signals. Of course, these are generally crude by scientific standards. However, we can imagine how to make them precise. With a clock, a pulsed light source, and a detector, we can transmit light pulses at some repetition rate and record the arrival times and frequencies of any signals that may be received. We can liken the returned signals to radarlike echoes from other observers and record them in terms of a space-time framework in each individual reference frame. Note that we have no metersticks or other devices to measure distances. All we measure are times. Nevertheless, within this framework we can determine distance, or, more precisely, measure the quantity we call "distance" in our model.

Consider two observers, Alf and Beth (see fig. 2.1). Assume that each has an accurate clock, calibrated against the standard atomic clock. Also suppose each has a device that can emit a sequence of electromagnetic pulses with a fixed time interval or *period* controlled by the ticks of their clocks. We can think of them as radar pulses, but, in principle, any part of the electromagnetic spectrum can be utilized. Further, suppose that Alf and Beth each have a device that will detect any electromagnetic pulses returned to them and can use the clock to measure the periods of the received pulses along with the time interval between the first pulse transmitted and the first returned. Let us assume, for illustration, they use *nanoseconds* (10^{-9} second) as their time unit.

In order to communicate, Alf and Beth must first synchronize their clocks. Alf sends out a series of pulses with equally spaced time intervals. Beth simply reflects back pulses at the same rate as she receives them, as shown in fig. 2.1(a). Alf then consults his clock. Suppose it reads the same time between the received pulses as those he transmitted. He defines this as meaning that they are "at rest" with respect to each other.

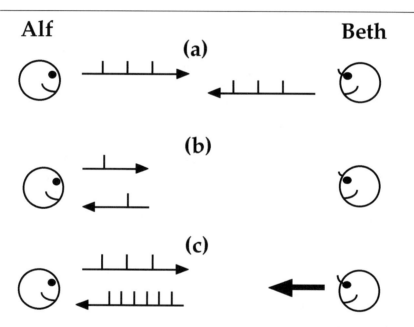

Fig. 2.1. In (a), Alf sends a series of pulses to Beth who reflects them back. Alf measures the same rate returned that he transmitted, so they are defined to be at rest with respect to each other. In (b), Alf sends a single pulse and measures the time it takes to return. This time interval is defined as twice the "spatial interval" between the two. In (C), Alf measures a shorter period for the return pulses and determines that Beth is approaching at some relative velocity.

Now suppose Alf sends out a single pulse and finds that its return is delayed by a certain amount, as seen in fig. 2.1(b). He then makes a model to describe this observation. Alf assumes that "out there" is an object, Beth, who reflected the pulses back to him. He imagines that a dimension called *space* separates Beth from him. He pictures the pulse as "traveling through space," taking a time equal to half the measured delay to reach Beth and another half to get back. The two are thus viewed in the model as being separated from one another by a "distance" equal to the half-time delay multiplied by an arbitrary constant c. The only purpose of c is to change the units of time from nanoseconds to meters or some other familiar unit of distance. Indeed, Alf can chose $c = 1$ and measure the distance in "light-nanoseconds" (about a foot). Beth does the same, so they agree on common time and distance scales. They call the speed of the pulses c the *speed of light.*

Note that the speed of light is *defined* in the model to be the same for both Alf and Beth. The distance from Alf to Beth is simply specified as the time it takes an electromagnetic pulse to go between the two, or half the time out and back. That spatial distance has no meaning independent of the time interval for a returned signal.

Experimenting with several observers, we can add further dimensions in space and introduce the concept of direction. Suppose we have one additional observer, and that all three can communicate with one another by suitably modulating the light pulses exchanged or exchanging additional electromagnetic signals. Then they each know their relative distances and can use them to form a triangle, as seen in fig. 2.2.

As a first approximation, let us use Euclidean geometry to describe the space, so the triangle is in a flat plane. The three distances are sufficient to determine the interior angles of the triangle using the familiar trigonometric law of cosines, and those angles can be used to specify a "direction" in space from one observer to another. The distances and directions can be used to define *displacement vectors* between the observers. A Cartesian coordinate system can also be introduced, as shown in the figure, which again we note can be located at any point and its axes aligned in any direction.

By *assuming Euclidean geometry* we have required that the sum of the interior angles of the triangle is 180°. We currently have the hindsight that Einstein formulated his theory of general relativity in terms of non-

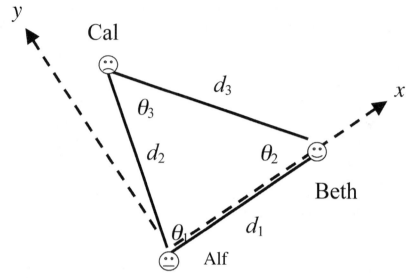

Fig. 2.2. The distances measured as described in the text can be used to form a triangle whose vertices specify the positions of each observer. The interior angles of the triangle can be determined from the distances using the law of cosines. Those angles can also be used to define the "directions" of the three displacement vectors between the positions of the observers. A two dimensional x, y coordinate system can be used to specify the positions.

Euclidean geometry, which is much more mathematically sophisticated and where the interior angles of triangles do not generally add to 180°. For now, however, let us stick to a Euclidean picture. The test, as always, is whether the model we develop agrees with the data. We will find that for most applications it does. When it does not, then we need another model for that data. Again, the operational nature of the process means that we avoid philosophical dispute about the "true geometry" of space. The geometry of space is what we choose based on the need to fit the data. Perhaps more than one choice will work.

Adding more observers, we find that a three-dimensional space is sufficient to describe most observations. Within that space, vector position displacements between each pair can be defined as having a magnitude equal to the distance between the observer positions and a direction that requires two angles, in general, which can be conveniently measured from some baseline. A Cartesian coordinate system can also be introduced with three coordinate axes and each position vector then represented by a set of three numbers, the components of the vector.

Again it must be emphasized that space is not necessarily "truly" three-dimensional. If we find in the future that more spatial dimensions are needed, as proposed by string theorists, then the model will be suitably modified.

Now suppose that on another occasion, Alf observes that the pulses reflected by Beth are received with a shorter period between pulses than those he transmitted, as in fig. 2.1(c). He models this by hypothesizing that Beth is "moving" toward him along their common line of sight. Alf can use the two periods to measure their relative velocity along that direction. If the return period is higher, Beth is moving away. In this manner, the concept of motion is introduced into the model. Note that it is *not* assumed beforehand. From the two periods, out and in, Alf can determine (measure) Beth's velocity as it pertains to the model.

This process is similar to how astronomers compute the velocity of galaxies, providing the primary evidence that the Universe is expanding (in their model). They observe that the spectral lines from most galaxies are shifted toward lower frequencies or higher periods compared to laboratory measurements. This *redshift* indicates the Universe is expanding as farther galaxies exhibit greater redshifts. Although they do not use radar-ranging in this case, the principle is the same.

We can extend the picture to three spatial dimensions by introducing a vector velocity—a set of three numbers whose components are the time rates of change of the position displacement vectors described above. Vector accelerations can also be measured as the time rates of change of velocities.

This should sound familiar to anyone who has studied elementary physics. However, note that the three-dimensional spatial picture is not presumed but derived from a series of measurements. The procedure is not totally arbitrary. We rely on quantitative observations and our space-time model must describe those observations. For example, we find that three dimensions are both necessary and sufficient for our spatial model. If we were simply dealing with a dream world, instead of reality, we would not find such limitations. So, while I keep emphasizing that we must not assume that reality is composed of the objects of our models, like space and time, those models are nevertheless constrained by whatever reality is out there.

I believe I have sufficiently emphasized the model-dependent nature of our physical concepts so that I need not always qualify my statements by "in the assumed model." However, this qualification should be understood throughout this book.

Now, you might ask, what if instead of electromagnetic pulses, which are composed of massless photons, Alf had sent out massive particles like golf balls? You would run into trouble with the distance scale. To see this, suppose Alf and Beth are moving apart and another observer, Cal, passes Beth as he is moving in the opposite direction toward Alf (see fig. 2.3). At the instant Beth and Cal pass one another, they each send a golf ball to Alf. The golf ball moves at the same speed in their reference frames. However, Beth's golf ball is moving more slowly with respect to Alf than Cal's, since she is moving away and Cal is moving toward Alf. Alf then measures different distances to Beth and Cal.

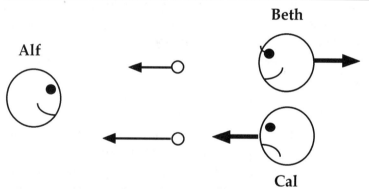

Fig. 2.3. Beth and Cal are at the same distance from Alf when they each send a golf ball to Alf. Since Beth is receding, her golf ball is moving more slowly than Cal's and so reaches Alf later. Alf concludes that Beth is farther away than Cal. This does not happen with photons since they always travel at the same speed.

From this, we conclude that the simplicity and consistency of our whole space-time model depends on the constancy of the speed of light. This is just what is assumed in Einstein's theory of special relativity. Special relativity follows automatically from his picture. Again, it need not have been the case that this model works. Not any arbitrary model will work—except in our dreams. But this model does work, and others, such as a golf ball model, do not (or, at least, cannot be made to work simply).

As we have seen, Milne's definition of space is now de facto and the constancy of the speed of light (or at least some universal speed c) is now deeply embedded in our definitions of space and time. If we insist on using meters for distance and seconds for time, the speed of light $c=3\times10^8$ meters per second. But, as we have seen, we could just as well have decided to work in units of light-nanoseconds for distance, in which case $c=1$. We should keep this in mind and think in terms of a world where $c=1$. This greatly simplifies our attempts to understand many relativistic ideas.

TOOLS

Any model we invent to describe the world must agree with the observations that we can make of that world. Let us proceed to review the tools humans have forged to describe the quantitative data they accumulate from experiments and observations.

In our space-time model of observations, distance and time are relational quantities. You measure the positions of a body at various times in relation to some other, arbitrary body. The position can be indicated by a vector whose length or "magnitude" is the distance of the body from some arbitrary reference point and whose direction is the angle with respect to some arbitrary reference line (see fig. 2.4).

Vectors are often described in the Cartesian coordinate system, where the projection of the vector on each of three mutually perpendicular axes gives the coordinates of the vector (see fig. 2.5). This tool is introduced early in the physics curriculum, not only because it provides a compact way to represent a quantity with magnitude and direction, but also because vectors do not depend on the coordinate system being used. You can translate the origin of that coordinate system to any other point in space, and rotate it by any desired angle, and the vector remains the same, as shown in fig. 2.6. We will see that this concept, which is called *invariance*, is the key ingredient in formulating the laws of physics. Those laws should not depend on the arbitrary choice of coordinate system.[6]

The time at which a body is at a specific position is similarly measured in relation to the time it was at another specified, arbitrary position.

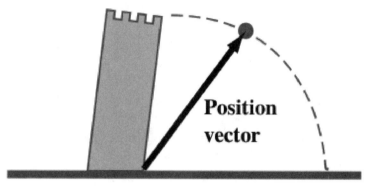

Fig. 2.4. The position vector of a particle. The magnitude of the vector equals its length. The direction of the vector can be specified by the angle above the horizon.

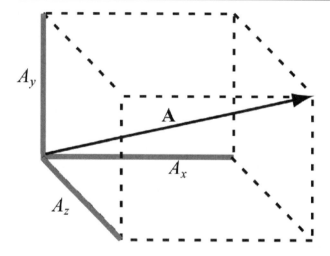

Fig. 2.5. A three-dimensional vector **A** can be projected against each of the three mutually perpendicular axes of a Cartesian coordinate system, giving the coordinates of the vector, A_x, A_y, A_z.

The origin used for the clock time, that is, the time you start the clock, is also arbitrary. From such measurements you can calculate the time rate of change of position, namely, the vector velocity, and the time rate of change of velocity, namely, the vector acceleration.

In relativistic physics, time is included with position to form a four-dimensional manifold called *space-time*. The "position" of a point in

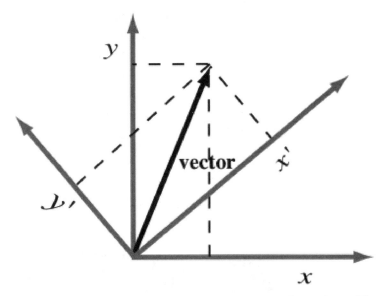

Fig. 2.6. The Cartesian coordinates of a vector in two dimensions. The magnitude and directions are unchanged to a translation or rotation of the coordinate system. That is, the vector is invariant to these transformations of the coordinate system.

space-time is thus a vector that specifies the vector position in familiar three-dimensional space, along with the time at which that position has been measured. Thus a point in space-time tells us not only *where* an event has happened, but also *when*. A simple Euclidean space-time does not agree with the data, so a "metric" must be used to define the type of space being utilized. In the simplest applications, a Euclidean space is used in which the additional dimension specifying a time t is taken to be ict, where $i = \sqrt{-1}$. In more advanced physics, non-Euclidean metrics are assumed with the time axis equal to the real number t.

MASS AND ENERGY

Mass is normally thought of as the amount of matter in a body, but this does not tell us much since we then need to first know what matter is. Instead, let us define *mass* as one of the quantitative measures of the inertia of a body, that is, the observed resistance of a body to changes in its condition of motion. The greater the mass of a body, the harder it is to start the body moving, to speed it up, to slow it down, or to stop it. The connection of mass to inertia has a long history and was made quantitative in Newton's laws of motion.

Consider an experiment in which you have two bodies at rest on a frictionless surface (or, say, far out in space). Place a coiled spring between the bodies and measure how much they accelerate after the spring is released (see fig. 2.7). Then the ratio of the masses is operationally defined as the inverse ratio of the recoil accelerations. If one of the masses is taken as a standard, say, one kilogram, then we can measure other masses in relation to this standard.

Fig. 2.7. Defining mass. The ratio of two masses is equal to the inverse ratio of their recoil accelerations.

This also gives us a provisional way to define force that will later be refined. If a body of constant mass m is observed to have a vector acceleration $\mathbf{a}$, then the net force on that body is the mass multiplied by the acceleration. This is the familiar form of Newton's second law of motion, $\mathbf{F} = m\mathbf{a}$. Note that the direction of the force is the same as the direction of the acceleration. I do not want to leave the impression that Newton's second law is an empty statement, a simple definition. It does, in fact, specify that a force is necessary for acceleration. However, as we will see, this is not a universal law.

Energy is a more recent physics concept that does not appear in early Newtonian physics. The English word *energy* comes from the Greek *energeia* for activity. Webster gives 1599 as the earliest date for the use of the term, but energy did not play an identifiable role in physics until as late as 1847. At that time, Hermann von Helmholtz (d. 1894) introduced the law of conservation of energy, which has proven to be one of the most powerful principles of physics. Let us list the various forms of energy that have been identified.

Kinetic energy is the quantity of energy associated with a body's motion. When a body is at rest, it has zero kinetic energy. The faster it is moving, the greater its kinetic energy.

Potential energy is stored energy that can be converted into other forms of energy. For example, a falling body exhibits the conversion of potential to kinetic energy as it falls. An arbitrary constant can always be added to potential energy without changing any results. Another way to say this is that only differences in potential energy matter. By convention, the gravitational potential energy between two bodies is negative.

Total mechanical energy is the sum of the kinetic and potential energies of a body.

Rest energy is a quantity that Einstein introduced in 1905 in his special theory of relativity. The rest energy E of a body of mass m is given by the familiar formula. Rest energy is a kind of potential energy that also can be converted into other forms of energy, usually kinetic energy, in chemical and nuclear reactions.

Einstein discovered an equivalence of mass and energy that was not previously realized. Mass can be used to produce energy and vice versa. Mass and rest energy are really aspects of the same entity, which is sometimes referred to as *mass-energy*. The total relativistic mass-energy of a body is, in general, the sum of its rest, kinetic, and potential energies.

In nontechnical literature you will often hear the term *relativistic mass*, which in relativity depends on the speed of a body. Here I follow the physics convention and use the term *mass* to refer to the rest mass. In the same regard, the term mass-energy introduced in the preceding paragraph should be understood to be shorthand for "rest mass-energy."

MOMENTUM

Another concept closely related to mass-energy is *momentum*, what Descartes and Newton identified as the "quantity of motion." Indeed, momentum captures more of the idea of inertia than mass-energy. A moving body has inertia in the form of momentum. The higher the momentum, the harder it is to change the motion of that body. More precisely, the momentum of a body moving at a low speed compared to the speed of light is the product of its mass and velocity. The direction of the momentum vector is the same as the direction of the velocity vector. Thus, the momentum of a body is just its velocity scaled by its mass. This formula must be modified for speeds near the speed of light.

The correct form of Newton's second law (which he used) is that the net force on a body is equal to the time rate of change of the momentum of that body, which reduces to the familiar $\mathbf{F} = m\mathbf{a}$ when the mass m is constant and the speed is much less than the speed of light.

RADIATION

Radiation is another term connected with the concept of mass-energy. Electromagnetic radiation is composed of particles called photons that have momentum and energy. Visible light is made of comparatively low-energy photons. Infrared and radio "waves" are composed of even lower-energy photons. Ultraviolet and x-ray photons are photons with higher energies than those in the visible spectrum. Gamma ray photons have higher energies still.

Nuclear radiation is also particulate in nature. It comes in three forms: alpha rays (α-rays), which are helium nuclei; beta rays (β-rays), which are electrons; and gamma rays (γ-rays), which are photons. Since gamma rays are photons, they travel at the speed of light. The electrons in beta rays generally move at "relativistic" speeds, that is, speeds near the speed of light. On the other hand, alpha rays, which also result from nuclear decay, are generally "nonrelativistic," that is, have speeds much less than the speed of light.

In this book, I will occasionally use the astronomical convention in which "matter" refers to bodies moving at nonrelativistic speeds and "radiation" refers to bodies moving at or near the speed of light (usually specifically photons). However, it should be understood that there is only one kind of "stuff" in the Universe with inertial properties, and that stuff is properly called matter. It is easily identifiable. When you kick matter (or radiation), it kicks back. Matter has quantitative properties that can be measured, called mass, energy, and momentum.

KICKING BACK IN MOMENTUM SPACE

Earlier I described the radarlike experiment in which we send and receive electromagnetic pulses, that is, pulses of photons, and use the relative timing of the incoming and outgoing signals to calculate the distance, velocity, and acceleration of the objects that we assume are out there in space reflecting the photons back to us. We saw that this picture of localized objects in space-time is a model we contrive to describe the measurements. We had no independent knowledge to confirm that those objects were "really out there," or that the quantities we calculated from clock measurements alone had some kind of concrete reality of their own.

To drive this point home, let us look at an alternative model that is not quite so intuitive but often turns out to be a more direct way to deal with the actual data. This model corresponds to one that is commonly used by elementary particle physicists in describing their experiments and in making computations using the model that describes these experiments. We will call this the *momentum space model*, where the use of "space" here refers not to familiar position space but to the more abstract use of the term that refers to any graphical description of variables.

The basic experiment that is done in particle accelerators involves taking a beam of particles of known mass and carefully calibrated energy and momentum, striking it against a target, and measuring the momenta and energies of the outgoing particles. In the earliest experiments, the target comprised a fixed block of matter. In more recent years, two beams have been brought into collision, greatly increasing the effective energy of the interaction. However, the basic processes involving particle collisions are generically equal and correspond to what we have taken as the archetype of human interaction with reality, kicking it and measuring the return kicks.

Experiments have been done with beams of protons, electrons, neutrinos, and other particles as well as photons. These are generalizations of the human observation process by which photons are reflected to our eyes as illustrated by the radar experiment. Only now, instead of a clock we will have a device that measures the color of each photon that we send out, and of each photon that is returned. By "color" in this context I mean the color of a light beam of identical photons. In fact, it corresponds to the energy of each photon in the beam. Measuring color gives us the photon energy, and also the magnitude of its momentum, which is simply equal to its energy divided by c.

Our eye-brain system comprises such a device. While we usually describe color measurements in terms of the wavelength of an electromagnetic wave, a *spatial concept*, we have known since the early twentieth century that a focused electromagnetic wave is a beam of photons that carries momentum and energy just like any other particle. A monochromatic beam, as from a laser, is monoenergetic.

Now we are in a position to describe our basic experiment in momentum space. The process is illustrated in fig. 2.8. The momentum of an incoming particle p_a corresponds to our kick, while the resulting momentum of the struck particle p_b corresponds to the kickback. The two particles need not be the same type. By making many measurements, we can obtain the probability distribution for different amounts of momentum transferred q from the kicker to the kickee.

The job of model building, then, is to find an expression for that probability to compare with experiment. This will involve some hypothesis about the nature of the object being struck. That hypothesis will be falsified by the failure to agree with experiments and (tentatively) verified by a successful agreement.

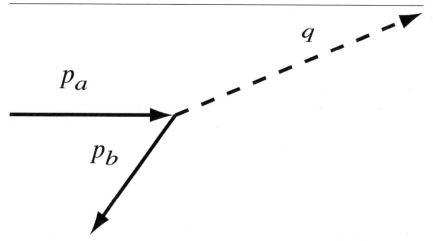

Fig. 2.8. The basic observation process in momentum space. An incoming particle of momentum p_a carries the "kick." A particle of momentum p_b carries the "kickback." The momentum transfer $q = p_a - p_b$. The arrows shown are graphically to scale.

Consider a particle interaction $a+c \rightarrow b+d$. In the scheme developed by Richard Feynman, interactions such as this are beautifully and intuitively illustrated by what are called *Feynman diagrams*. As shown in fig. 2.9, the momentum transfer is pictured as being carried by an unobserved or "virtual" particle X that is emitted by particle a, which is then absorbed by the target particle c. Feynman's methods allow for the calculation of a quantity called the *probability amplitude*, which is in general a complex number. In the simplest case, this amplitude is simply a function of the momentum transfer carried by X and some constants that measure the strength of the interaction at the two interaction points, or *vertices*, in the diagram. In general, more than one particle can be exchanged, and many other diagrams must be considered. In principle, we must sum the amplitudes of all Feynman diagrams with the same incoming and outgoing particles but with different numbers of internal interactions. The number of terms in the sum is infinite but a good approximation is obtained by considering only the simpler ones. The probability distribution is then the square of the absolute value of that sum. The interference effects that

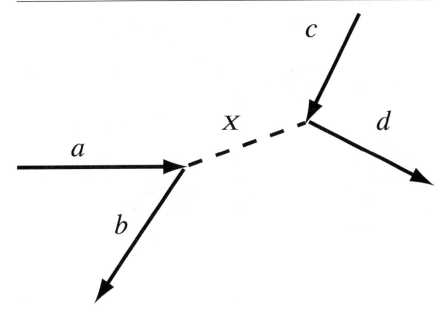

Fig. 2.9. Feynman diagram for the process $a+c \rightarrow b+d$. In the theory, an undetectable or "virtual" particle X is exchanged, which can be viewed as carrying the momentum transferred to c.

typify quantum phenomena and are usually associated with the "wave nature" of particles, then, arise from the different phases of the amplitudes being summed—with no mention of waves.[7]

One specific example of this is *Rutherford scattering*, where *a* and *b* are alpha particles, *c* is the struck nucleus, and *d* is the recoil nucleus. An alphaparticle is the nucleus of the helium atom, which is spinless. In this case, the exchanged particle *X* is the photon. In the approximation where the single photon exchange diagram dominates, an amplitude is calculated which agrees with the experiment. Rutherford scattering was the process by which Lord Rutherford inferred the nuclear model of the atom.

NOTES

1. James Boswell, *Life of Johnson* (1791. Oxford: Oxford University Press, 1980), p. 333.

2. J. Butterfield and C. J. Isham, "Space-time and the Philosophical Challenge of Quantum Gravity," in *Physics Meets Philosophy at the Planck Scale*, ed. C. Callender and N. Huggett (Cambridge: Cambridge University Press, 2001).

3. See http://www.physics.nist.gov/GenInt/Time/boulder.html.

4. See http://tf.nist.gov/cesium/fountain.htm.

5. E. A. Milne, *Relativity, Gravitation, and World Structure* (Oxford: Oxford University Press, 1935).

6. While the terms *velocity* and *speed* are often used interchangeably in the vernacular, in physics speed is more precisely defined as the magnitude of the velocity vector.

7. Most physicists will tell you that the Feynman diagram is just a tool to aid in writing down the equations needed to make calculations. However, in my book *Timeless Reality* (Amherst, NY: Prometheus Books, 2000), I show how it is possible to reify the diagrams, that is, picture them as describing real events, by utilizing time reversibility.

CHAPTER THREE

POINT-OF-VIEW INVARIANCE

THE COSMOLOGICAL PRINCIPLE

The models of physics are designed to describe observations. However, they are not simply photographs of nature taken from various angles. If that were the case, we would need a different model—a new photograph—for every situation. That would not be wrong, but the number of such models in that case would be infinite and of little practical use. Instead, physicists seek *universality*, formulating their laws so they apply widely and do not depend on the point of view of any particular observer. In that way, they can at least hope to approach an accurate representation of the objective reality that they assume lies beyond the perceptions of any single individual.

This requirement was not always so obvious. Self-centeredness had led humans for thousands of years to picture the Universe as revolving around them. Even one of the greatest thinkers of all time, Aristotle, described motion in terms of a fixed Earth, the abode of humanity, located at the center of an absolute space. When Nicolaus Copernicus (d. 1543) challenged that doctrine and suggested that Earth moved around the Sun, he set in motion a chain of events that would lead to the seventeenth-century scientific revolution. This revolution began in earnest when Galileo Galilei (d. 1642) looked through his telescope and convinced himself that Copernicus was correct. In fact, he said, Earth moves.

The notion that no center of the Universe exists is called the *Copernican principle*, although technically Copernicus had simply removed Earth from the center and replaced it with the Sun. With the enormous expansion in our knowledge of the Universe in the years since Copernicus, the Copernican principle has been made universal and to include time as well as space. Let me express it as follows:

The cosmological principle: Physical models cannot depend on any particular position in space or moment in time.

That is, when we formulate our models, they should not contain explicit references to specific points in space and time.

This does not say that the Universe looks exactly the same at every point in space and time. Obviously, it does not. The view from the top of the Rockies, near where I now live, is not at all like the view from Waikiki, near where I used to live. And the view from Waikiki, at least along the shoreline, is not the same now as it was when I first stood on that beach in 1963, when the tallest building was the five-story Royal Hawaiian hotel. On the astronomical scale, telescopic images provide a vast range of vistas and good evidence that those vistas have altered dramatically over time.

Whether an earthly panorama of mountains and oceans or deep-space photographs of distant galaxies, observations indicate that the same physical models can be used as far out in space and as far back in time as our instruments take us. That is to say, while all phenomena may not look the same in detail, they can be modeled in terms of the same underlying principles. For example, the spectral lines of hydrogen bear the same relation to one another when measured for a quasar billions of light-years away, and thus billions of years in the past, as they do when measured in a laboratory on Earth. They are shifted in wavelength, but this is taken into account by postulating that the Universe is expanding.

INVARIANCE, SYMMETRY, AND CONSERVATION

In chapter 2 I noted that vectors are used in physics in order to express equations, such as those describing the motions of bodies, in a way that does not depend on any particular coordinate system. Those equations are then *invariant* to a change in the coordinate system, such as when the origin is moved to another point or the coordinate axes are rotated.

This invariance is precisely what is required by the cosmological principle. That principle furthermore requires that our equations not depend on the particular time that we start our clock. The physics equations describing motion taught at the most elementary level already implicitly implement three invariance principles, which I will state as follows:

Space-translation invariance: The models of physics are invariant to the translation of the origin of a spatial coordinate system.

Space-rotation invariance: The models of physics are invariant to the rotation of the axes of a spatial coordinate system.

Time-translation invariance: The models of physics are invariant to the translation of the origin of the time variable.

Coordinates are still useful for recording the positions of a body at different times, but the particular coordinate system used is simply chosen for convenience. For example, if you are on top of a building and will be dropping a ball down to me at the base of the building, your preference might be to call your own position and the initial position of the ball $y = 0$. And you might start your stopwatch, that is, define $t = 0$, when you release the ball. On the other hand, I might prefer to call my position and the final position of the ball $y' = 0$ and already have my stopwatch running when you drop the ball. The phenomenon that is observed is the same in either case. The same distance is covered in the same time interval in both cases. So, it clearly makes no sense to formulate our model describing the experiment in such a way that depends on one particular point of view. This suggests that we can make the following, further generalized principle:

Point-of-view invariance: The models of physics cannot depend on any particular point of view.

This includes the three space-time invariance principles stated above, and will be generalized further to the more abstract representations of observations that we use in modern physics.

However, in this chapter we will limit ourselves to the consequences of point-of-view invariance in space and time.

Often invariance is expressed by the term *symmetry*. For example, a sphere is invariant to rotations about any axis and so possesses what we call *spherical symmetry*. A cylinder is invariant to rotations about a single axis and has cylindrical symmetry. A snowflake has rotational symmetry in steps of 60 degrees about one axis. This symmetry is discrete as compared to the continuous symmetry of a cylinder. The above space-time invariance principles are also referred to as space-time symmetries.

In 1915 mathematician Emmy Noether proved that certain mathematical quantities called the *generators* of continuous space-time transformations are conserved when those transformations leave the system unchanged.[1] Furthermore, she identified the generators with energy, linear momentum, and angular momentum. This was a profound find, implying that the great "laws" in which these three quantities are conserved are simple consequences of the symmetries of space and time. Only in recent years has Noether gained deserved recognition for her profound achievement.[2] The implications can be summarized as follows:

- In any space-time model possessing time-translation invariance, energy must be conserved.
- In any space-time model possessing space-translation invariance, linear momentum must be conserved.
- In any space-time model possessing space-rotation invariance, angular momentum must be conserved.

Thus, the conservation principles follow from point-of-view invariance. If you wish to build a model using space and time as a framework, and you formulate that model so as to be space-time symmetric, then that model will automatically contain what are usually regarded as the three most important "laws" of physics, the three conservation principles. As we will now see, further symmetries will require the introduction of other "laws," not to govern the behavior of matter, but to govern the behavior of physicists so that when they build their models, those models will be point-of-view invariant.

GALILEO'S PRINCIPLE OF RELATIVITY

As we saw in chapter 1, Galileo realized that, contrary to common intuition, we cannot distinguish when we are "moving" and when we are "at rest." Otherwise, how can you explain the fact that we do not sense our motion as we ride Earth at 30 kilometers per second around the Sun?[3] Suppose we are in a closed chamber, unable to see outside. We cannot tell whether we are at rest or moving at constant velocity. A good example is our experience when riding in a modern jetliner. We have no sense of the high speed at which we are moving relative to the ground. What motions we do notice are changes in velocity—accelerations—as the aircraft speeds up, slows down, or changes direction momentarily in turbulent air. In practice, this means that if we attempt to write down mathematical relationships to describe our observations, these cannot contain an absolute velocity.

The Principle of Galilean relativity: Physical models cannot depend on absolute velocity.

To obey Galileo's principle of relativity, a mathematical model of physics must apply in all coordinate systems that are moving at constant velocities with respect to one another. We see this is a further generalization of the space-time symmetries discussed above in which the coordinate systems were at rest with respect to one another and we simply changed their origins and orientations. Now those coordinate systems can have a relative velocity.

At this point, let me begin to use the common term *reference frame* to indicate the space-time point of view of an observer.

A mathematical operation called the *Galilean transformation* converts any equation written on one reference frame to the corresponding equation applying another frame moving at a constant relative velocity. Since physics equations refer to quantities measured with metersticks, clocks, and other instruments, measurements in a given reference frame are done with instruments that are fixed in that frame. The Galilean transformation relates the two sets of measurements. When the form of an equation is unchanged under a Galilean transformation, we say that it is *Galilean invariant*. The principle of relativity requires that all the equations of physics be Galilean invariant. We see that this is a natural extension of point-of-view invariance to moving reference frames.

Let us apply the principle of relativity to the famous Galileo experiment in which cannon balls are dropped from the Leaning Tower of Pisa. To an observer on the ground, a ball dropped from the top of the tower falls along a straight line to the ground, as shown in fig. 3.1(a).

Now consider another observer riding on a boat that is moving in at constant velocity in the river that runs through Pisa near where Galileo is making his observations (before the view was obscured by buildings). That observer sees the object fall along a parabolic path, as shown in fig. 3.1(b). If we take the equation that describes the path of a falling body

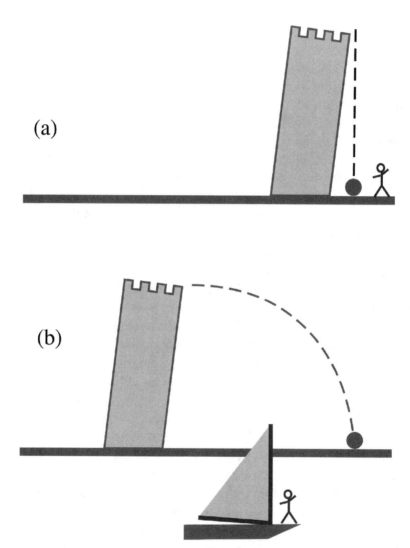

Fig. 3.1. In (a), an observer standing next to the tower sees an object dropped from the tower fall in a straight line. In (b), an observer on a boat moving in a nearby river sees the object fall in a parabolic arc. For clarity, the tower is shown only at its position when the object is released. In (b) it will move horizontally so the object will still land at the foot of the tower.

in (a) and apply the Galilean transformation, we will get the path observed in (b).

So, it would not be point-of-view invariant to state that bodies dropped from above fall in a straight line to the ground. They fall along parabolas to some observers. More accurately, bodies fall along a parabolic trajectory in all reference frames, where a straight line is a special case of a parabola.

In 1905 Einstein showed that the Galilean transformation must be modified for relative speeds near the speed of light c and replaced by the *Lorentz transformation*. This becomes the Galilean transformation in the limit of speeds low compared to c.

For many centuries, people believed that a force was necessary to produce motion, as taught by Aristotle. This suggests what might be called "Aristotle's law of motion," that force depends on velocity. However, such a law violates Galilean invariance. There can be no distinguishable difference between being at rest and being in motion at constant velocity.

Newton, following Galileo, proposed instead that force depends on *changes* in velocity, or, more precisely, changes in momentum. Since the time rate of change in velocity or momentum is the same in all reference frames moving at constant velocity with respect to one another (assuming speeds much less than the speed of light), this "law of motion," Newton's second law, is Galilean invariant.

Let us play a little game. Suppose you did not know about Newton's second law and were given the following assignment: Without relying on any observational data, write down the simplest equation for the force that is proportional to the mass of the body (assumed constant), and a polynomial containing its position, velocity, acceleration, the time rate of change of acceleration, the time rate of change of that, and so on. The equation must be invariant under space translation, space rotation, time translation, and a Galilean transformation. Assume speeds much less than c and, for simplicity, let us just work in one dimension.

From space-translation invariance, the force cannot depend on absolute position. As we have seen, velocity is not Galilean invariant, so terms with velocity also must be excluded. The acceleration, however, is invariant under all the assumed operations, so we can have terms proportional to the acceleration and any power of the acceleration, in all possible combinations. The simplest, however, is just a single term directly proportional to the acceleration, $\mathbf{F}=m\mathbf{a}$. Furthermore, the time rate of change of the acceleration, the time rate of change of that, and so on would still obey all the invariance principles and could have, in principle, been necessary to describe motion. But the data show that they also are not needed. The world seems to be as simple as it can be.

Of course, much of what we see in the Universe does not look all that simple. But I will argue that this complexity can still be understood in the simplest of terms.

FICTITIOUS FORCES

In Newtonian physics, we find an important apparent exception to the second law of motion. Think of what happens when you are driving a car with left-hand drive making a right turn around a corner. Viewed from outside, in a reference frame attached to the road, the car door is pressing against you with a force, called the *centripetal force*, which has the effect of changing the direction of your velocity vector. The time rate of that velocity vector change equals your acceleration. Note that even if your speed is constant, your velocity changes when its direction changes. This is standard Newtonian physics, with $\mathbf{F}=m\mathbf{a}$ still applying.

Now, let us look at what is happening as viewed from your reference frame inside the car. Your body is pressed against the door but no apparent force is pushing you from the right. Your acceleration in the car's reference frame is zero, yet you have the force of the door acting on your left side. This *violates* Newton's second law, $\mathbf{F}=m\mathbf{a}$.

So, we seem to have two kinds of reference frames. Those points of view in which the second law of motion is obeyed are called *inertial reference frames*. Those points of view in which the second law of motion is violated are called *noninertial reference frames*.

However, we can still utilize the second law of motion in non-inertial reference frames by introducing what are commonly referred to as *fictitious forces*. The centrifugal force is one such fictitious force, which we imagine acting to balance forces in a rotating reference frame, such as inside a turning car. It comes into play for bodies on the surface of a rotating body such as Earth, as does the equally fictitious Coriolis force.

Introducing these fictitious forces allows us to apply the second law on the surface of the rotating Earth, a handy thing for Earth inhabitants. Note that fictitious forces need not be introduced in the inertial reference frame of an observer sitting above Earth and watching it rotate below her.

In chapter 1 we saw that Newton's laws of motion followed from momentum conservation, with force defined as the time rate of change of momentum. In short, we find that all of Newtonian mechanics is a consequence of point-of-view invariance.

SPECIAL RELATIVITY

Maxwell's equations of electrodynamics are not Galilean invariant. In the late nineteenth century, Hendrick Lorentz found that these equations are invariant under a different transformation we now know as the *Lorentz transformation.*

The Lorentz transformation, as applied by Einstein in his 1905 theory of special relativity, entails that distance intervals measured with a meterstick and the time intervals measured with a clock will be, in general, different for two observers moving relative to one another. In particular, moving clocks appear to slow down (*time dilation*) and a moving body appears to contract along its direction of motion (*Lorentz-Fitzgerald contraction*).

Einstein derived the Lorentz transformation from the assumption that the speed of light is absolute and otherwise maintained the principle of relativity. The result was a rewriting of some of the most basic mathematical expressions of mechanics. Einstein showed that the old expressions are applicable only as approximations that apply when the speeds involved are much less than the speed of light.

In special relativity, the Lorentz transformation replaces the Galilean transformation. (Note that the latter is a special case of the former for relative speeds low compared to the speed of light). The change from Galilean to Lorentz transformation does not affect the other space-time symmetries and their corresponding conservation laws.

The question still remains: What is the nature of light if it is not an etheric vibration? Einstein answered that in another paper appearing the same year, 1905: Light is composed of particles, just as Newton originally thought. We now call those particles *photons*. However, we need to use Einstein's revised equations rather than those of standard Newtonian physics to describe the motion of photons.

Let us consider the implications of the fact that distance and time intervals, as measured with metersticks and clocks, are relative. For example, if we could build a spacecraft that was able to leave Earth at a constant acceleration equal to one *g*, the acceleration of gravity on Earth (providing artificial gravity to the crew), it could travel to the neighboring galaxy, Andromeda, which is 2.4 million light-years away (accelerating the first half of the trip and then decelerating the second half), in only thirty years elapsed time as measured on the clocks aboard the ship. If the ship turns around, the astronauts will return home aged by sixty years, while almost 5 million years will have passed on Earth.

The relativity of space and time is supported by a century of experiments that confirm the equations of special relativity to high precision. This fact is not easily reconciled with a model of objective realty that contains space and time as substantial elements. On the other hand, relativity conforms nicely to a model of reality in which space and time are simply human contrivances, quantities measured with metersticks and clocks. This is not to say that these concepts are arbitrary; they help describe objective observations that presumably reflect an underlying reality. However, we should not simply assume that those observations coincide with that reality—whatever it may be.

In 1907 Hermann Minkowski helped place special relativity on an elegant mathematical foundation by introducing the notion of four-dimensional *space-time*, where time is included as one of the four coordinates. This neatly accommodates the fact that in order to define the position of an event you need three numbers to specify its position in familiar three-dimensional space and another number to specify the event's time of occurrence.[4] This is an extension of Newtonian space-time, with the addition of four-dimensional rotations.

The Lorentz transformation can be shown to be equivalent to a rotation in Minkowski space. It follows that Lorentz invariance is equivalent to rotational invariance in space-time. Thus, special relativity also follows from point-of-view invariance. We can even introduce a quantity that is analogous to angular momentum, which is conserved as a consequence of spatial rotation symmetry. However, that quantity is not particularly useful.

We can further generalize the cosmological principle to include Lorentz invariance in the following way:

The principle of covariance: The models of physics cannot depend on our choice of reference frame.[5]

Here "reference frame" refers to a four-dimensional space-time coordinate system. The principle of covariance tells us that we must formulate our model so that it does not single out the origin of that coordinate system or the direction of its axes. Conservation of energy, linear momentum, angular momentum, and all of special relativity follow. Of course, this is again just point-of-view invariance.

GENERAL RELATIVITY

Aristotle supposed the existence of an absolute space with respect to which everything moves. Galileo showed that this was not supported by the observable fact that motion at constant velocity is indistinguishable from being at rest. Newton incorporated Galileo's principle of relativity in his laws of motion; however, he assumed that an absolute reference frame nevertheless existed with respect to which bodies accelerate. Leibniz objected, arguing that other bodies were necessary to define motion, that space and time are relational concepts.

We can see the difference between Newton's and Leibniz's views by considering a spinning disk, like the record on an old-fashioned phonograph turntable. If we crumple up a small bit of paper and place it on the edge, it will fly off when the disk reaches some critical rotational speed. Now, suppose that the disk and paper are all by themselves in the Universe. According to Newton, the paper will still fly off since the disk is rotating with respect to "absolute space." Leibniz, on the other hand, would say that the paper remains in place since the disk cannot rotate without some other object to provide a reference. Unfortunately, we can never do this experiment to test which of the great men, Newton or Leibniz, was correct on this issue.

In the late nineteenth century, Ernst Mach provided further insight into this dispute when he suggested that the inertia of matter is a consequence of all the other matter in the Universe. *Mach's principle*, as it is known, was never formulated as a precise mathematical statement.[6] The main point seems to be that a body sitting all by itself in the Universe would have zero inertia, which would seem to agree with Leibniz. In our example, in an otherwise empty universe the crumpled paper would remain on the disc since it has no inertia (mass).

For our purposes here, I will adopt a simple notion, which one might attribute to Leibniz or Mach:

Leibniz-Mach principle: In order for a body to accelerate there must exist at least one other body.

Once again, this is a rather obvious requirement of the operational nature of space, time, and the other quantities we form from measurements. If we have only a single body, we have no way to measure acceleration. So, acceleration is meaningless in that case.

While Einstein said he was influenced by Mach, you will not find a direct application of Mach's principle in his published papers or typical textbook derivations of general relativity. However, Einstein did apply another notion that lay largely unspoken in Newton's laws of motion and gravity—the equivalence of gravitational and inertial mass. Recall that the second law of motion equates the net force on a body to the product of its mass and acceleration, in the case where the mass is constant. This mass is called the *inertial mass*. On the other hand, mass also appears in Newton's law of gravity. This mass is called the *gravitational mass*. Technically, they need not be equal since they are defined by different observations. But the data indicate that they are the same. This leads to the following principle:

The principle of equivalence: The inertial and gravitational masses of a body are equal.

This principle, which has now been checked to extraordinary precision, implies that one cannot distinguish between the effect of acceleration and a gravitational field, at least in a small, localized region of space.

Consider fig. 3.2. An astronaut is inside a capsule in outer space, far from any planet or star. The capsule is accelerating with respect to an inertial reference frame. The astronaut stands on a bathroom scale that measures his "weight" to be a product of his mass and the acceleration of the capsule. It is as if he were on a planet where the acceleration of gravity, g, was equal to the acceleration of his capsule, a. For example, if the capsule's acceleration were equal to $g=9.8$ meters per second per second, he would think (to a first approximation) that he was sitting in a

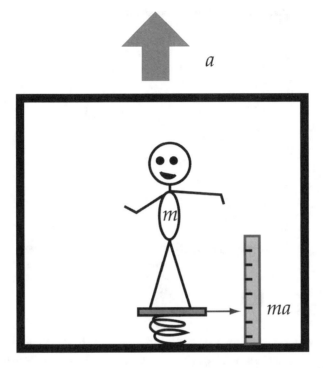

Fig. 3.2. An observer in an accelerating capsule in outer space will experience a weight.

closed room on Earth. However, since he is, in fact, far from any planet, his weight is a result of acceleration rather than gravity.

Einstein describes sitting in his chair in the patent office in Bern in 1907 and suddenly realizing that a person in free fall would not feel his own weight. As illustrated in fig. 3.3, it is as if the person were floating out in space far from any other bodies, such as stars or planets.

Now, Einstein understood that a uniform gravitational field is not easily attainable, requiring an infinitely large body. Field lines converge toward the center of gravity of any finite gravitating body and this is, in principle, detectable in a falling elevator. So the principle of equivalence is a *local* rather than a *global* principle, strictly applying to infinitesimal regions of space.

Let us see how the Leibniz-Mach and equivalence principles lead to general relativity. Consider a freely falling body, as in fig. 3.3. Since the body does not change position in its own reference frame, both its

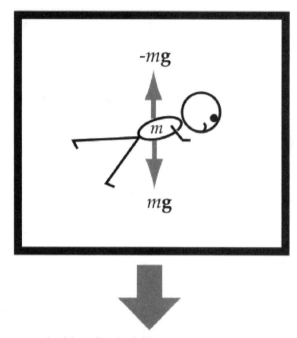

Fig. 3.3. A person inside a freely falling elevator has zero acceleration in his own reference frame. Assuming he knows that he is on Earth, he must introduce a force to balance the weight. Alternatively, he might simply assume there are no forces acting on him, including gravity.

velocity and its acceleration are zero in this reference frame. A freely falling body experiences no net external force. Therefore, whatever equations we use to describe gravity must reflect the fact that an observer in a closed capsule in free fall cannot detect whether he is accelerating or sense the presence of gravity.

Next, let us consider a second reference frame fixed to a second body, such as Earth. An observer on Earth witnesses a body accelerating toward Earth. She then introduces a "force of gravity" to act to accelerate the body.

In the Newtonian picture, a gravitational "field" can be calculated for any known distribution of gravitational masses. In the case of a point mass, we get the familiar inverse square law for the magnitude of the field, which, by the equivalence principle, equals the acceleration of the body. The direction of the gravitational field vector at any point in space points to the mass.

However, the Newtonian relationship is not Lorentz invariant. By the principle of equivalence the gravitational mass equals the inertial mass, and by special relativity the inertial mass equals the energy (divided by c^2). The energy density will vary between reference frames. In the place of energy (inertial, gravitational mass) density as the source of gravity, Einstein used the *momentum-energy tensor*, also known as the *stress-energy tensor*, to represent the distribution of matter in a covariant, Lorentz-invariant way. We can think of this quantity as a 4×4 matrix or table of sixteen numbers. One entry in the table is the energy density. The other components comprise energy and momentum flows in various directions.

Einstein proposed that the gravitational field is proportional to the momentum-energy tensor, with the constant of proportionality chosen to give the Newtonian form in the nonrelativistic limit. Since the energy-momentum tensor is Lorentz invariant, the gravitational field is also a Lorentz-invariant tensor.

In what has become the standard model of general relativity (other models exist), Einstein related the gravitational field tensor at a space-time point to the curvature of space-time at that same point in a non-Euclidean geometry. This assured that all bodies would follow a *geodesic* in space-time in the absence of forces, the gravitational force thus being eliminated and replaced by a curved path. An example of a geodesic is a great circle on the surface of a sphere, a circle that passes through the sphere's center. If you are traveling between two points on the surface, and can't burrow through the sphere, then the shortest distance is along a great circle.

Included in Einstein's final equation was a quantity Λ, the infamous *cosmological constant*. At the time Einstein published general relativity, the prevailing belief held that the Universe was in static equilibrium. Although planets and stars moved with respect to one another, the Universe was assumed to be, on average, a "firmament" of fixed and perhaps infinite volume. In both Newton's and Einstein's theories of gravitation, however, a firmament is unstable. Einstein found that he could add a constant repulsive term in his equation to balance the normal gravitational attraction of matter.

It is often reported in the media and in many popular books on cosmology that the cosmological constant was a "fudge factor" introduced by Einstein to make things come out the way he wanted. Perhaps that was his motivation, but the fact is that unless one makes further assumptions, a cosmological constant is required by Einstein's equations of general relativity and should be kept in the equations until some principle is found that shows it to be zero.[7] When Einstein later heard of Hubble's discovery that the Universe is expanding, he called the cosmological term "my biggest blunder." For many years the measurements of the cosmological constant gave zero within measuring errors, but in the past two decades Einstein's "blunder" has resurfaced again in modern cosmology.

Recall that Newtonian physics was left with the problem of noninertial reference frames in which Newton's second law of motion is ostensibly violated. It was also left with the problem of understanding the nature of gravity. In Newtonian physics, gravity is a mysterious, invisible force that acts instantaneously over great distances and causes a body to accelerate. Einstein produced a single solution for both problems. He noted that it was impossible to locally distinguish gravity from acceleration, so gravity is a force introduced to explain the fact that a body with no visible forces acting on it is observed to be accelerating in some reference frames. For example, we see a body accelerate from the top of a tower to the ground. We can see no forces acting on that body, although we can infer an upward force from air friction. We want to believe Newton's second law, which says that acceleration requires a force. So, we introduce the force of "gravity" to save Newton's law. Since a body obviously does not accelerate with respect to itself, then no net force on a body exists in the body's own reference frame.

By representing inertia in terms of the Lorentz-invariant energy-momentum tensor, Einstein guaranteed that his theory made no distinction between reference frames. The actual equations that relate inertia to the geometry of space are largely determined by covariance and, in fact, represent the simplest form of those equations consistent with this requirement and the Leibniz-Mach and equivalence principles.

The great success of general relativity[8] and its specific formulation in terms of non-Euclidian space-time appears, at first glance, to support the notion that space-time constitutes the fundamental stuff of the Universe. Einstein seems to have adopted that view, which is still widespread among physicists today.[9] However, many different mathematical procedures can model the same empirical results. One can find alternate descriptions of general relativity in which Euclidean space is maintained.[10]

Most physicists do not believe that general relativity will be the final word on gravity, despite close to a century of successful application. General relativity is not a quantum theory and at some point effects of quantum gravity are expected to be exhibited. However, after forty years of effort along several different lines of attack, a quantum theory of gravity is still out of reach.[11] No doubt the reason can be traced to the complete lack of empirical guidance. One possibility, hardly ever mentioned, is that no such theory will ever be developed because it has no empirical consequences that can ever be tested. But that remains to be seen.

In any case, we have witnessed another example of how physicists are highly constrained in the way they may formulate the laws of physics, and when they properly take into account those constraints, the laws fall out naturally. The equations that physicists use to describe observational data should not be written in such a way as to single out any particular space-time coordinate system or reference frame. The principle of covariance generalizes earlier notions, such as the Copernican and cosmological principles and the principle of Galilean relativity. I have also termed it point-of-view invariance. The application of this principle is not a matter of choice. Centuries of observations have shown that to do otherwise produces calculations that disagree with the data.

NOTES

1. Nina Byers, "E. Noether's Discovery of the Deep Connection between Symmetries and Conservations Laws," *Israel Mathematical Conference Proceedings* 12 (1999), http://www.physics.ucla.edu/~cwp/articles/noether.asg/noether.html (accessed November 5, 2005). This article contains links to Noether's original paper, including an English translation.

2. Leon M. Lederman and Christopher T. Hill, *Symmetry and the Beautiful Universe* (Amherst, NY: Prometheus Books, 2004), pp. 23–25, 69–77.

3. Technically, Earth is not moving in a straight line so it has a small acceleration that could, in principle, be detected. But to a first approximation, our high velocity is unnoticeable.

4. Four-dimensional space-time is not Euclidean, although it can be made so by using an imaginary time axis.

5. For a good discussion see Richard C. Tolman, *Relativity, Thermodynamics, and Cosmology* (1934. Mineola, NY: Dover, 1987).

6. J. Barbour and H. Pfister, eds., *Mach's Principle—From Newton's Bucket to Quantum Gravity* (Boston: Birkhauser, 1995).

7. Steven Weinberg, *Gravitation and Cosmology: Principles and Applications of the General Theory of Relativity* (New York: Wiley, 1972).

8. For a discussion of the empirical tests of general relativity see Clifford M.Will, *Was Einstein Right? Putting Relativity to the Test* (New York: Basic Books, 1986).

9. But see J. Butterfield and C. J. Isham, "Space-time and the Philosophical Challenge of Quantum Gravity," in *Physics Meets Philosophy at the Planck Scale*, ed. C. Callender and N. Huggett (Cambridge: Cambridge University Press, 2001).

10. R. H. Dicke, *Theoretical Significance of Experimental Relativity* (New York: Gordon & Breach, 1964).

11. For recent reviews, see Lee Smolin, *Three Roads to Quantum Gravity* (New York: Basic Books, 2001); Butterfield and Isham, "Space-time and the Philosophical Challenge of Quantum Gravity."

CHAPTER FOUR

GAUGING THE LAWS OF PHYSICS

THE STATE OF A SYSTEM

In classical physics, the state of a particle at any specific time is given by the position and momentum of the particle at that time. More complex bodies can have additional "internal" states that result, for example, from rotation or vibration. But the classical pointlike particle lacks any of these internal states.

We can represent the classical state of a system of particles as a point in *phase space*, an abstract space in which the axes are the spatial coordinates and momentum components of the particles. This is illustrated in fig. 4.1 for a single particle moving in one dimension. Given the net force on a particle, the equations of motion predict the particle's trajectory through phase space, that is, how the particle's position and momentum vary with time.

In general, at least a six-dimensional phase space is needed to describe the motion for each particle. A system of N particles requires $6N$ dimensions. More dimensions are needed for internal degrees of freedom, such as spin. Obviously these are impossible to illustrate graphically in most cases, but the mathematical description is a straightforward extension of the familiar three-dimensional description.

In quantum mechanics, a point in an abstract, multidimensional space, called *Hilbert space*, mathematically represents the state of a system. We locate that point with a vector ψ, which we will call the *state vector*. Let us call our abstract space ψ-*space*. I need not get specific at this point about the structure of this abstract space. Indeed, we need not

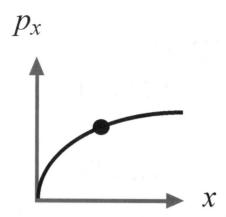

Fig. 4.1. In classical mechanics, the state of a particle is indicated by its position in phase space. Its motion can be described as a trajectory in phase space. Here a single particle is moving in one dimension. In general, if we have N particles moving in three position-coordinate dimensions, phase space has $6N$ dimensions.

even limit our current discussion to the formal state vector of quantum mechanics. In principle, any set of observables or mathematical functions of the observables used to describe some system can be thought of as "coordinates" of the state space of that system.

For example, consider a vastly oversimplified model of an economic system in which we have just two parameters, *Supply* and *Demand*, defining the state of the economy. These parameters can form the coordinates of a two-dimensional economic state space, with a point in that space representing the state of the economy at a given time, as shown in fig. 4.2(a). A "state vector" is drawn pointing from the origin of the coordinate system to that point. An economic model would then predict how that state vector depends on supply and demand.

GAUGE TRANSFORMATIONS AND THEIR GENERATORS

For our physical application, let us start simply with an abstract state space of only two dimensions. One convenient way to represent this space is as a "complex" plane (Argand diagram), as illustrated in fig.

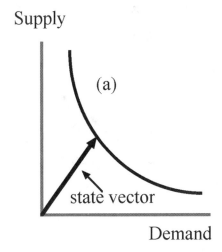

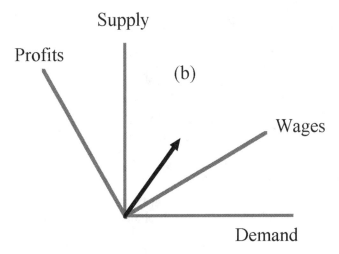

Fig. 4.2. In a simple, two-parameter model, a vector in a two-dimensional abstract space can represent the state of an economy. An economic theory would predict how the state depends on the parameters. In (a) we show Adam's model in which the axes are Supply and Demand. In (b), Karl's model the axes are Profits and Wages have been added. This is viewed as a rotation of the coordinate axes. Gauge symmetry applies when the state vector is invariant to the coordinate transformation, meaning the two models are simply two different points of view of the same state and thus equivalent.

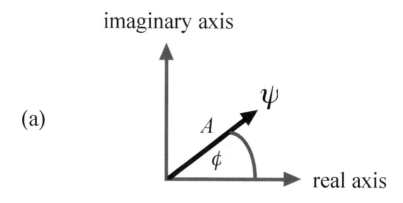

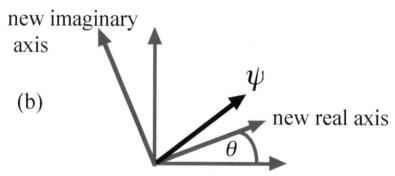

Fig. 4.3. The state vector ψ in two-dimensional complex space. In (a) the magnitude A and phase ϕ are shown. In (b) we see that ψ is invariant under rotations of the complex plane, that is, it still has the same magnitude and direction, although the phase is changed. This is an example of gauge invariance.

4.3(a). In that case, the state vector is a single *complex number*, ψ. We will call this plane "ψ-space."

The algebra of complex numbers is a very valuable mathematical tool and is used throughout physics. A complex number c is any number of the form $c = a + ib$, where $i = \sqrt{-1}$; a is called the *real* part and b is the *imaginary* part of c.

So, we can think of the state vector ψ as a two-dimensional vector, whose projection on one coordinate axis is the real part of ψ and whose projection on the other axis is the imaginary part. As shown in the figure, ψ can also be specified by a length or *magnitude A* and *phase ϕ*.

Now, let us rotate the coordinate system through some angle θ, as shown in fig. 4.3(b). We will call that rotation a *gauge transformation.* Note that, while the phase of ψ is changed by a gauge transformation, the state vector itself remains fixed in magnitude and direction. We term this property *gauge symmetry* or *gauge invariance.*

When θ is the same for all points in space-time, we have a *global* gauge transformation. When θ varies with spatial position or time, we have a *local* gauge transformation.

When ψ is a vector in higher dimensions, θ is represented by a matrix rather than a single number. That matrix is a table of numbers needed to describe the rotation. The basic idea of a gauge transformation as analogous to a rotation in an abstract state vector space is maintained and gauge symmetry is viewed as invariance under such rotations. In general, we call θ the *generator* of the transformation.

Now, the choice of coordinate system in ψ-space is analogous to the choice of coordinate system in position space. It is arbitrary, depending on the point of view of the person making the choice. It thus seems reasonable to adopt the same principle here that we did in the previous chapter for the space-time representation of particle motion. That is, we postulate that in order to describe objectively the state of a system in ψ-space, the description must be point-of-view invariant. In this case, that means gauge invariant.

The principle of gauge invariance: The models of physics cannot depend on the choice of coordinate axes in ψ-space.

Recall our earlier economics example (fig. 4.2). Suppose we had one economist, Adam, whose model used the parameters Supply and Demand and a second economist, Karl, whose model used two other parameters, Profits and Wages. A gauge transformation would take us from one model to the other, as a rotation of coordinate axes from Supply-Demand to Profits-Wages, illustrated in fig. 4.2(b). Gauge invariance would hold if the two models described the economic facts equally well.

Notice an important point here. The models need not have described the facts equally well. This demonstrates why the process is not arbitrary, not just a matter of cultural choice as postmodernists might claim. Ultimately, the data decide.

CONSEQUENCES OF GAUGE INVARIANCE

Let us now examine the consequences of gauge invariance. Consider the motion of a particle in space-time. Suppose we move or "translate" the time axis by an infinitesimal amount, that is, start our clock an instant earlier or later. This amounts to a change in the representation of the state of the particle, that is, a rotation of the coordinate system in ψ-space. Gauge invariance requires that the state of the system does not change under such an operation.

Recall that one of the elements of the cosmological principle is time translation invariance. We can also translate and rotate the spatial axes and similarly apply gauge invariance. It can be shown mathematically that the generator of the time translation operation is just the energy of the particle. Similarly, the generators of the translations along the three spatial axes are the components of the linear momentum of the particle along those axes. And the generators of the rotations around the three spatial axes are the components of the particle's angular momentum around those axes.

We now see that gauge invariance is a generalization of space-time covariance. That is, we can regard gauge invariance as the general principle and space-time covariance as a special case of gauge invariance. We can anticipate other conservation principles might arise as we describe observables in an abstract, multidimensional framework that goes beyond space and time. Indeed, in the case of a charged particle the generator of the rotation of the state vector in complex space, that is, a phase change, is just the *electric charge* of the particle. Thus, *charge conservation* is a consequence of gauge invariance. In modern particle physics, where the state-vector space is multidimensional, other conservation principles arise by the same mechanism.

Let us recall another significant consequence of space-time symmetries. In special relativity we think of an event as occurring at a specific point in a suitably defined four-dimensional space-time.[1] Just as we can rotate the spatial coordinate axes in three-dimensional space, we can also rotate the space-time coordinate axes in four dimensions. When we do this, we find that the process is precisely the Lorentz transformation used by Einstein in special relativity. It then follows that the principles of special relativity, including the constancy of the speed of light in a vacuum and $E=mc^2$, are a direct consequence of gauge symmetry.

In short, we have shown that the great conservation principles of classical physics (energy, linear momentum, angular momentum, charge) and the principles of special relativity, which includes the principle of Galilean relativity and the constancy of the speed of light, are nothing more than rules forced upon us by point-of-view invariance, expressed as gauge invariance.

Furthermore, as we saw in chapter 2, Newton's laws of motion follow from momentum conservation; so these are also a consequence of point-of-view invariance. As I have mentioned, fluid mechanics and statistical mechanics can be derived from Newton's laws, and classical thermodynamics can be derived from statistical mechanics. In chapter 3 we saw how gravity is represented in general relativity as the fictitious force needed to preserve Lorentz invariance when transforming from one reference frame to another accelerating relative to the first. Newton's law of gravity follows as a special case.

This covers all of classical physics except electrodynamics, which is codified in Maxwell's equations. Amazingly, these too can be derived from point-of-view invariance. Although I have introduced the gauge transformation in the language of quantum state vectors, none of the above discussion assumes any quantum mechanics. Indeed, gauge invariance first appeared in classical electrodynamics.

Neither the classical nor the quantum equations of motion for an electrically charged particle are invariant to a local gauge transformation. However, let us postulate that local gauge invariance also must be maintained in order to achieve point-of-view invariance. When we insist on this, we find that we need to introduce additional forces—much as we did for gravity. These forces are described by vector fields that are identical to the electric and magnetic fields of Maxwell's equations!

And so, all the fundamental laws and principles of classical physics, plus special and general relativity, are seen as simply arising from the requirement that our descriptions of reality must be formulated so as to be independent of any particular point of view. If we did not include any of these "laws" in our physical models, we would not be describing reality in a way that is independent of point of view.

GAUGED QUANTUM MECHANICS

This leaves quantum mechanics and the later developments of twentieth-century physics, especially the standard model of particles and forces. When the mathematical formalism of quantum mechanics is made gauge invariant, the laws of quantum mechanics, including the Heisenberg uncertainty principle, follow. The mathematical procedures of gauge models lead directly to the mathematical descriptions used in quantum mechanics. Let me try to give some idea of how this happens.

We have seen that the generators of gauge transformations correspond to common physical quantities, such as energy and momentum, and these are conserved under gauge invariance. However, measured physical quantities are simple, real numbers, while, mathematically, the generators of transformations are operators that do not always obey the same arithmetic rules as numbers.

Consider the operators R_x, R_y, and R_z that rotate a coordinate system (x, y, z) about its respective coordinate axes. For simplicity, let us look at the case where each angle of rotation is 90 degrees.

In fig. 4.4(a), the effect of applying R_z first, followed by R_x, is shown. In (b), we apply R_x first, followed by R_z. Note that we get different final axis orientations in the two cases. You can illustrate this result with your hand, forming three mutually perpendicular axes with your first three fingers. We conclude that the operator product $R_z R_x$ is not equal to the operator product taken in the reverse order, $R_x R_z$. That is, operators do *not* in general obey the *commutation rule* of familiar numbers, as, for example, $4 \times 2 = 2 \times 4$.

So, in general we cannot use familiar numbers to represent operators in our model, although the measurements themselves are numbers. In quantum mechanics, these numbers are associated with the *eigenvalues* of the corresponding operators.

When Heisenberg developed the first formulation of quantum mechanics in 1925,[2] he proposed that matrices, which are tables of numbers, represent observables. Matrices can be used to represent the rotation operators described above. The commutation rule does not apply to matrix multiplication, making them suitable, but not unique, for this purpose. Heisenberg's matrices obeyed the classical equations of motion, but did not in general commute.

Heisenberg's matrix mechanics is not normally taught in elementary

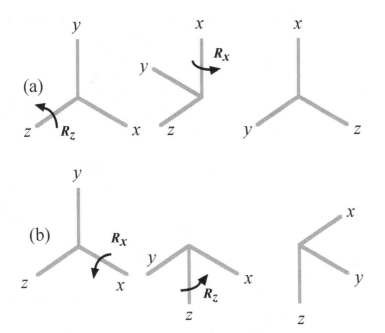

Fig. 4.4. The noncommutativity of rotation operators. In (a) the coordinate system is first rotated around the z-axis then around the x-axis. In (b) the order is reversed, with different results. Thus, $R_x R_z \neq R_z R_x$.

quantum mechanics courses because of the use of the more advanced mathematics of matrix algebra. Instead, quantum mechanics is usually introduced in the form developed by Erwin Schrödinger a year after Heisenberg, in 1926. In Schrödinger's quantum mechanics, observables are represented by the differential operators of more familiar calculus, which also do not in general commute. Still, the best way to think of observables represented in quantum mechanics is as operators that rotate coordinate systems in ψ-space, the way R_x, R_y, and R_z rotate coordinates in position-vector space. Modern quantum mechanics, based on the work of Dirac, is done in this way.

THE UNCERTAINTY PRINCIPLE

In the most familiar example, the operator for the component of momentum along a certain axis and the corresponding position coordinate do not commute. We can sort of see how this happens if we think about measuring the position x and momentum p_x of a particle. Suppose we measure x and get a value x_o. Suppose that in another experimental run, we measure p_x before we measure x. Recall that p_x is the generator of a translation along the x-axis. Let a be the amount we move the axis. Now, when we measure x we will get the value $x_o - a$. So, as with the rotation operators, the order of operation of certain pairs of quantities, called *conjugate* variables, is important.

The commutation rule for x and p_x can be derived directly from gauge ideas. And this, in turn, by a mathematical theorem, leads to the Heisenberg uncertainty principle. Commutation rules can be obtained for other observables as well. In general, an uncertainty principle exists for all pairs of observables, conjugate variables whose operators do not commute. Pairs that do commute, such as y and p_x, can be measured simultaneously with unlimited precision.

The uncertainty principle is almost obvious, once you think about it. As I have emphasized, matter is what kicks back when you kick it. In order to observe an object of matter, you must interact with that object. At some level, when the object is small enough, that interaction is bound to interfere with what is being observed. We simply did not notice this effect right away, because it is generally negligible on the everyday scale. A pebble does not recoil when we shine a flashlight on it. On the atomic and subatomic scales, however, the interaction between observer and observed becomes unavoidable. It is more like looking at a pebble with a machine gun.

An important result of deriving quantum mechanics from gauge invariance is that Planck's constant h, which appears in the uncertainty principle, is shown to be an arbitrary constant—just as we found for the speed of light c. Planck's constant can have any value except zero. If it were zero, then there would be no uncertainty principle, no limit to how accurate we can measure positions and momenta simultaneously.

This unique result is not found in conventional treatments of quantum mechanics and might come as a surprise to some readers, even those who may be physicists. The quantity $h/2\pi$ is usually designated by the symbol $\hbar$, called "hbar." In what physicists call "natural units," $\hbar = c = 1$.

TIME EVOLUTION

Gauge invariance also provides for a simple derivation of one of the most basic equations of quantum mechanics, the *time-dependent Schrödinger equation*. (The *time-independent Schrödinger equation* is a special case). This is the equation of motion of quantum mechanics, which specifies how the state of a system evolves with time. In the gauge picture, this is simply the time translation operation. The energy operator, called the *Hamiltonian*, is the generator of this operation. That is, time evolution is a gauge transformation.

THE SUPERPOSITION PRINCIPLE

Some of the postulates made in deriving quantum mechanics are assumptions about the mathematical procedures to be used, and not statements about what is "really" going on. For example, the quantum wave function (a way to represent the state vector) is defined as a complex number whose squared magnitude gives the probability per unit volume for finding a particle in an infinitesimal volume centered at a particular point in space. Is this a law of physics? I would not think so, unless the wave function is itself an element of reality, which is by no means an accepted fact. Rather, this is the definition of a mathematical quantity in the model that we will use to calculate probabilities.[3] Just as we introduce space and time to describe motion, we introduce the wave function to describe a quantum state. Both are ingredients in the models that we use to describe observations. The test, as always, is whether they do so or not.

Still, it would be nice to have some reason beyond simply that "it works" to hypothesize the connection between the wave function and probability. The interference effects that mark the difference between quantum and classical particle physics arise in the model from that source. Indeed, since classical waves exhibit interference effects, the common explanation used in elementary discussions is that a "duality" exists on which particles are sometime waves and waves sometimes particles.

However, this duality is not present in the basic model, nor does it properly describe observations. Even in an experiment such as double-slit interference, whether done with light or electrons, sufficiently sensitive detectors will pick up individual, localized particles. In the case of light, which we normally think of as waves, the particles are photons. What is observed is a statistical distribution, or intensity pattern, of particles that can be mathematically described in terms of waves. It is hard to see what it is that is doing the waving.

So, the wave function is used to calculate probabilities. As I have noted, at a more advanced level, the quantum state is represented by a state vector in an abstract vector space. This vector space is assumed to be "linear" in the same way that our familiar three-dimensional space is linear. In the latter case, we can always write a vector as the linear combination of three vectors, called *eigenvectors*, each pointing along a particular coordinate axis. The same is assumed for state vectors, with the generalization applying to any number of dimensions.

Now, we have seen that gauge invariance requires that the state vector be independent of the coordinate system. Thus it can be written as the superposition of any set of eigenvectors, each set corresponding to a different coordinate system. This can be expressed as the following principle:

The superposition principle: The state vector is a linear combination of eigenvectors independent of the coordinate system defining the eigenvectors.

The superposition principle is what gives us quantum interference and the so-called entanglement of quantum states. Each eigenvector represents a possible state of the system, and so quantum states are themselves superpositions, or coherent mixtures, of other states.

This would not be the case for nonlinear vectors. So, once again we find that a basic principle of physics, the superposition principle, is needed to guarantee point-of-view invariance, as is the linearity of the vector space.

While the mathematical formalism of quantum mechanics has proved immensely successful in making predictions to compare with experiment, no consensus has yet been achieved on what quantum mechanics "really means," that is, what it implies about the nature of reality. Numerous philosophical interpretations can be found in the literature, all of which are consistent with the data; any that were inconsistent would not be published. No test has been found for distinguishing between these interpretations (see my *The Unconscious Quantum* for further discussion).[4]

ANGULAR MOMENTUM AND SPIN

We usually associate angular momentum with the motion of a particle moving in an orbit about some focal point; for example, a planet moving in an elliptical path around the Sun.[5] This is referred to as the *orbital angular momentum.* In quantum mechanics, the orbital angular momentum of a particle is described by a quantum number that is always an integer or zero. This quantum number is approximately the magnitude of the angular momentum in units of $\hbar$.

A rotating body, such as Earth, possesses an angular momentum about its axis of rotation. This is the sum of the angular momentum of all the particles inside Earth, which are orbiting in circles about the Earth's axis. This total angular momentum is called *spin.* On the quantum scale, you would expect the spin quantum number of a rotating body to be an integer, since it is the sum of orbital angular momenta.

Early in the history of quantum mechanics, it was discovered that particles such as the electron, which otherwise appeared pointlike, possess properties, such as magnetism, that imply an intrinsic spin. In classical physics, a spinning electric charge has a magnetic field. However, a pointlike classical particle has zero spin and zero magnetic field. The fact that point particles possess spin has been, from the beginning, a great puzzle since it has no classical analogue. What is it that is spinning? Furthermore, the quantum number that is associated with the spin of the electron was found to be $s = 1/2$ and so could not be the net effect of total orbital angular momenta within a finite sized the electron. Angular momentum is the generator of spatial rotations. The corresponding operators are related to the rotation operators described above. Recall that the rotation operators for each of the three axes do not mutually commute, which implies that the corresponding angular momentum operators also do not commute. This means that only one component of angular momentum can be measured with unlimited precision at a given time.

The commutation rules for angular momentum operators can be derived from the rotation operators. They then can be used to show that the quantum number that is associated with the total angular momentum can be an integer or half-integer.

This is one of a number of unique phenomena that occur only at the quantum level and cannot be explained by models built on everyday experience. However, in this book I have developed the notion that whatever models physicists assemble to describe their data must be point-of-view invariant if they are to describe objective reality. Clearly a pointlike particle sitting all by itself in space is invariant under spatial rotations. It follows that conserved quantities will exist that correspond to the generators of rotations about three axes. Those generators are the components of angular momentum; the mathematical properties of those generators imply that a particle will have an intrinsic angular momentum, or spin, that can be zero, an integer, or a half-integer. This may not be a satisfying explanation of spin for those looking for familiar pictures in their models, but we see that spin is a necessary consequence of point-of-view invariance.

Particles with zero or integer spin are called *bosons*. Particles with half-integer spin are called *fermions*. The photon is a spin-1 boson. The electron is a spin-1/2 fermion. Fermions and bosons behave very differently. Two fermions cannot exist in the same state, according to the *Pauli exclusion principle*, which can be derived from basic quantum mechanics. Exclusion is confirmed by the very existence of complex chemical elements, as listed in the periodic table. By contrast, bosons like to be in the same state, as exemplified by *bose condensates*. The vacuum itself can be a bose condensate, filled with photons and other bosons of zero energy.

THE STANDARD MODEL

The full application of gauge invariance came into its own in the 1970s with the development of the standard model of elementary particles and forces. This model contains a common, if not totally unified, picture of the electromagnetic, strong, and weak forces. Gravity is not included. As discussed above, Maxwell's equations, which fully describe classical electrodynamics, follow from gauge invariance. This derivation can be performed in the quantum framework and leads to the same conclusion, that the electric and magnetic forces are introduced to maintain local gauge invariance. Quantum electrodynamics (QED) is a gauge model.

I do not want to leave the impression that gauge invariance is all that there is to quantum electrodynamics. Heroic mathematical efforts by

Julian Schwinger, Sin-Itiro Tomanaga, Richard Feynman, and Freeman Dyson in the late 1940s were needed to develop QED into a workable model that made calculations to compare with the data. They succeeded spectacularly, but this story has been told many times and so it need not be repeated here.[6]

The success of QED in providing a powerful model for calculating the electromagnetic interactions between particles, such as photons and electrons, was not followed immediately by similar success for the other forces. However, after two decades of steady progress in both model building and experiment, models of the weak and strong forces were developed and, joined with QED, now form the standard model.

Once again, the notion of gauge invariance provided the foundation for these models. The weak and strong nuclear forces were introduced into the mathematics in order to preserve local gauge invariance. In the 1970s Steven Weinberg, Abdus Salam, Sheldon Glashow, and others developed the *electroweak model* in which electromagnetism and the weak force were combined in a single framework. The generators for the gauge transformation in this case are three 2×2 matrices. The three conserved quantities corresponding to these generators are the components of an abstract three-dimensional vector called the *weak isospin* that is analogous to particle spin, which we have seen is already a more abstract concept than simple angular momentum.

The electroweak model made several predictions that were soon confirmed by experiments. Most important, the model predicted the existence of four new particles, the *weak bosons*, W^+, W°, W^-, and Z, which mediate the weak force much as the photon mediates the electromagnetic force. These particles were found with exactly the expected masses.

Another group of physicists worked out a gauge model of the strong force, called *quantum chromodynamics* (QCD). The generators of the gauge transformation in this case are eight 3×3 matrices. The corresponding conserved quantities are called *color*, by analogy, with the familiar primary colors red, green, and blue. These colors are carried by quarks. Antiquarks have the complementary colors: cyan, magenta, and yellow. The proton and neutron, which form the constituents of the nuclei of atoms, are composed of three quarks, one of each primary color adding up to white. Other particles called *mesons* are composed of quark-antiquark pairs that also add up to white (red + cyan = white, etc.). That is, all directly observed particles are "colorless." For decades physicists have searched for free particles of color and never found one.

Strongly interacting particles called *hadrons* are composed of quarks. The proton and neutron are hadrons. Additionally, hundreds of very short-lived hadrons exist. Most are *baryons*, like the proton and neutron, which are made of three quarks. Others are mesons, like the *pi meson* or *pion*, which are made of quark-antiquark pairs. Recently, new baryons comprised of five quarks have been discovered. In this case, a quark-antiquark pair supplements the three quarks that normally constitute baryons. They, too, are colorless.

According to QCD, the strong force is mediated by the exchange of massless particles called *gluons*, which exist in eight varieties.

While QCD has not had the astounding predictive success of the electroweak model, applications have been in wholesale agreement with the data.

NOTES

1. By "suitably defined" I mean that the geometry ("metric") has to be such that the position four-vector is invariant.

2. Although quantum theory begins with Max Planck in 1900 and his notion of the quantum of energy, the earliest quantum models, such as those of Bohr, were quite ad hoc. Heisenberg, in 1925, and then Schrödinger, independently a year later, made the first attempts at a fairly complete theory based on a few postulates.

3. David Deutsch claims to prove that the probability axioms of quantum mechanics follow from its other axioms and the principles of decision theory. See his "Quantum Theory of Probability and Decisions," *Proceedings of the Royal Society* A456 (2000): 1759–74.

4. Victor Stenger, *The Unconscious Quantum: Metaphysics in Modern Physics and Cosmology* (Amherst, NY: Prometheus Books, 1995).

5. Technically, even a particle moving in a straight line will have a nonzero angular momentum about any point not along that line.

6. For a complete history of QED, see S. S. Schweber, *QED and the Men Who Made It: Feynman, Schwinger, and Tomonaga* (Princeton, NJ: Princeton University Press, 1994).

CHAPTER FIVE

FORCES AND BROKEN SYMMETRIES

FIELDS AND QUANTA

We have seen that in order to preserve local gauge symmetry we must introduce a vector field into the equation of motion for a charged particle. That field is exactly the electromagnetic potential that follows from Maxwell's equations. Thus, analogous to gravity, the electromagnetic force is needed to preserve point-of-view invariance.

In quantum field theory, each field is associated with a particle called the *quantum* of the field. For the electromagnetic field, the corresponding quantum is the photon. Local gauge invariance also can be used to show that the photon is massless.

The generator of the gauge transformation in this case is a single number, the electric charge. The equation of motion for charged particles is invariant to global gauge transformations, which leads to the principle of charge conservation.

Recall that generators are in general mathematical operators that can be represented as matrices. In this case, we can trivially think of the charge, a single number, as a 1×1 matrix. A state vector that is invariant under the operations of the set of 1×1 matrices is said, in the language of group theory, to possess U(1) *symmetry.*

Group theory is the branch of mathematics that deals with transformations. Since numbers commute, this case is an example of what is called *abelian symmetry.*

The fields corresponding to the weak nuclear force appear when the equation of motion is also required to possess local gauge symmetry under the operation of a set of three 2×2 matrices. These constitute the generators of rotations in an abstract three-dimensional space called weak isospin space. They form the group of transformations called SU(2). Since they are generators of rotations, they do not commute and symmetry under SU(2) is thus *nonabelian*. Weak isospin is mathematically analogous to angular momentum, the components of which form the generators of rotations in familiar three-dimensional space. The three components of weak isospin are conserved since the equation of motion is invariant to SU(2) global gauge transformations.

Four vector fields are introduced to preserve local SU(2) symmetry; their corresponding quanta are four weak bosons: W^+, W^o, W^-, and B. If the gauge invariance in this case were perfect, these would be massless quanta, like the photon. However, at the "low" energies of current experimentation, the local gauge symmetry is "broken" (the global symmetries are okay) and the weak bosons are surprisingly heavy, eighty to ninety times more massive than protons. Furthermore, the neutral bosons W^o and B mix together to produce the observed particles—the photon and Z boson. Broken symmetry is an important part of our story, and will be discussed in detail below.

In quantum chromodynamics, a quantity called *color* is conserved by global gauge invariance under transformations of the state vector in eight-dimensional "color space." The transformation group here is called SU(3). The strong nuclear force appears when the equations are required to possess local gauge symmetry in this abstract space. Eight vector fields are introduced to preserve that symmetry; their quanta are eight *gluons*. Since the symmetry is unbroken, the gluons are massless.

These gauge symmetries form the essential ingredients of the standard model of elementary particles and fields that was developed in the 1970s.[1] The fundamental particles that comprise the model are shown in fig. 5.1. I must make it clear, however, that while some of these particles, like the photon, follow from local gauge invariance, we cannot currently derive the whole scheme from a single symmetry principle. Indeed, the fact that we have so many particles in the table indicates that the underlying symmetries are broken at our level of observation. So, the standard model is still part theory and part experiment, what is called an *effective theory*.

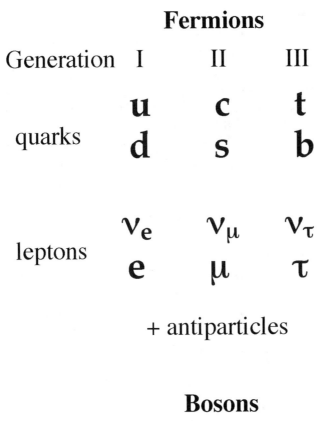

Fermions

Generation I   II   III

	I	II	III
quarks	u d	c s	t b
leptons	ν_e e	ν_μ μ	ν_τ τ

+ antiparticles

Bosons

W+

γ W⁰   Z

W-

+ 8 gluons

Fig. 5.1. The fundamental particles of the standard model of particle physics, grouped as Fermions (spin-1/2) and Bosons (spin-1). The three columns of fermions show the particles of each generation, with increasing masses to the right. Fermions constitute familiar matter while bosons mediate the forces between them.

Nevertheless, the standard model has remained fully consistent with all measurements made at the highest-energy particle accelerators to the date of this writing. The model had to be slightly modified with the discovery in 1998 that neutrinos have a tiny mass (in an experiment in which I participated), but this did not represent a paradigm shift. Nothing in the original formulation required that the neutrinos be massless and zero mass had been assumed initially for simplicity.

The specific ways in which broken symmetries appear in the standard model are currently not derived from fundamental principles but formulated to agree with the data. Sometime during the first decade of the new millennium, experiments at high-energy laboratories are expected to reveal physics beyond the standard model, that is, the more basic physics out of which the standard model arises.

It should not be surprising that our low-energy observations do not reflect the underlying symmetries that probably existed at the much higher energies of the early Universe. After all, our point of view is limited by the capabilities of our instruments and the fact that we live at a time when the Universe has cooled to only a few degrees above absolute zero. The complex structure of our current Universe most likely froze out of a much simpler state, as a snowflake freezes out of a featureless region of water vapor.

THE BASIC INTERACTION

The interactions between quarks and leptons are viewed in terms of exchanges of particles by way of the familiar Feynman diagrams. Some examples are shown in fig. 5.2. In (a), we have two electrons scattering from one another by the exchange of a photon. In (b), an electron and a positron collide and annihilate into a photon, which then produces another electron-positron pair. These are typical electromagnetic interactions and were already well described by the predecessor of the standard model, quantum electrodynamics. In (c), we have the prototype weak interaction in which a d quark decays into a u quark and a W-boson, and the W then decays into an electron and an antielectron neutrino. If we imagine this process as accompanied by another u and d quark that do not participate in the interaction, we have neutron beta-decay. In (d), we have a typical

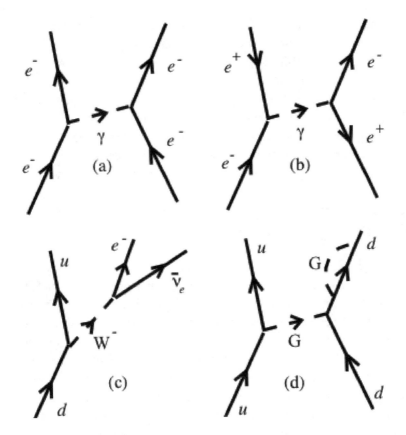

Fig. 5.2. Four examples of Feynman diagrams. See the text for details.

strong interaction in which a *u* and *d* quark exchange a gluon. Also shown is another gluon being emitted and then reabsorbed by the outgoing *d* quark. This sort of "radiative correction" can happen in all processes.

In general, a quark or lepton is pictured as emitting a boson, which carries energy and momentum across space, and then is absorbed by another quark or lepton. More than one particle, including quarks and leptons, can be exchanged. The elementary particle events recorded by detector systems result from many Feynman diagrams with multiple boson and fermion exchanges, "loops," and other complications.

Despite this, the basic interaction between elementary particles can be viewed simply as three lines connecting at a point or vertex. Each line represents the momentum and energy (4-momentum) of a particle that can either go into the vertex or come out. The total momentum, energy, charge, and other quantities are conserved at a vertex.

For example, consider fig. 5.3. The three-line interaction $e^+ + e^- \to \gamma$ (a) can also be written $e^- \to e^- + \gamma$ (b), $\gamma + e^+ \to e^+$ (c), and $\gamma \to e^+ + e^-$ (d). When you change the direction of a line, you move the particle to the other side of the reaction, being sure to reverse its momentum and energy and change the particle to its antiparticle. If it has "handedness" (technically called *chirality*), then in general you have to change that as well, for example, left into right. Handedness for a particle occurs when it is spinning along or opposite its direction of motion. Think of pointing your thumb in the direction of motion and curling your fingers in the direction of the spin. There are two possibilities, depending on whether you use your right or left hand.

The property that allows us to switch particles around in this fashion is called *CPT invariance*, where C is the operator that changes a particle to its antiparticle, P is the operator (called *parity*) that changes a particle's handedness (as if viewing the particle in a mirror), and T is the time-reversal operator. Most physical processes are invariant to each of the operations individually, but, as we will see below, a few rare processes are not. No violations of the combined operation *CPT* have ever been observed, and this can be viewed as another form of point-of-view invariance.

So, we can view the basic interaction as $f + \bar{f} \leftrightarrow b$ where f is a fermion (quark or lepton), $\bar{f}$ is an antifermion, and b is a boson. Or, equivalently, we can write $f \leftrightarrow f + b$. The vertex then represents a point in space-time where the interaction takes place. All the interactions of elementary particles can be assembled from this basic three-line interaction. For example, the electromagnetic interaction is $e^+ + e^- \leftrightarrow \gamma$, where the electrons can be replaced by *muons*, $\mu^+ + \mu^- \leftrightarrow \gamma$, or *tauons*, $\tau^+ + \tau^- \leftrightarrow \gamma$. Generically we can write these $\ell^+ + \ell^- \leftrightarrow \gamma$, where ℓ is any of the three leptons. Quarks can also interact electromagnetically: $q + \bar{q} \leftrightarrow \gamma$.

Similarly, the strong interaction is $q + \bar{q} \leftrightarrow G$, where q is a quark, $\bar{q}$ an antiquark, and G is a gluon.

The weak interaction can involve quarks and leptons:

$$q + \bar{q} \leftrightarrow Z \quad \ell + \bar{\ell} \leftrightarrow Z \quad q_1 + q_2 \leftrightarrow W \quad \ell_1 + \bar{\ell}_2 \leftrightarrow W$$

and so on.

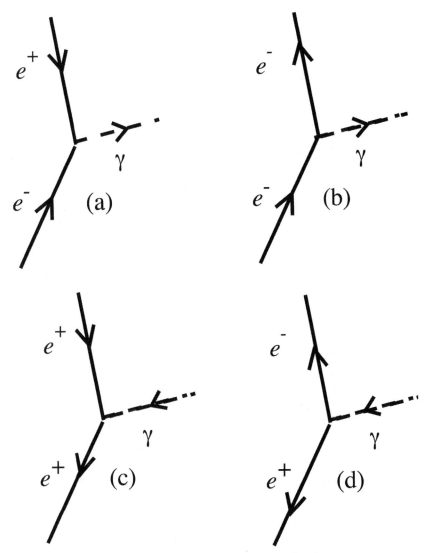

Fig. 5.3. The basic three-line interaction illustrated for the case $e + e \leftrightarrow \gamma$.

The rules for computing the probabilities for elementary particle processes using Feynman diagrams were first developed by Richard Feynman and Freeman Dyson in the 1950s. These can be derived from the basic gauge-invariant equations of motion. The calculated probabilities can be related to experimental measurements, such as scattering cross sections or particle decay rates.

Many technical problems had to be worked out when these methods were first introduced. The initial, highly successful application to purely electromagnetic processes in quantum electrodynamics (basically electron-photon interactions) was not followed immediately by similar successes in the weak and strong interactions. However, with the development of the standard model in the 1970s, these interactions are now incorporated into a single scheme. Gravity is the only known force that has not been successfully integrated, but this represents only a theoretical problem, since quantum effects of gravity are still unobservable.

The calculations are relatively easy when only one or two vertices contribute and are carried out routinely by theoretical physicists. In principle, you have to add together all the possible diagrams, an impossible task. By good fortune, however, a few diagrams suffice in many important cases.

THE RANGES OF FORCES

The electromagnetic force acts over great distances. Light has been seen from quasars over 10 billion light-years away. Each photon from the quasar is like the exchanged photon in fig. 5.2(a), where the emitting electron resides in the quasar and the absorbing electron resides in the astronomical detector on Earth.

By contrast, the nuclear forces act only over ranges on the order of the size of a nucleus, about 10^{-15} meter. How can we understand such a vast difference?

Although the notion of photon exchange was not fully developed until after World War II, the idea had been kicking around in the 1930s. In 1934 Hideki Yukawa suggested, by analogy, that the strong nuclear force was mediated by the exchange of an undiscovered particle. He showed that the range of forces, described in this fashion, would be proportional to the inverse of the mass of the exchanged particle. Since the mass of the photon is zero, the range of the electromagnetic force is infinite. The short range of the nuclear force resulted from the exchanged particle having mass.

Yukawa called his particle the *mesotron*. He estimated its mass to be intermediate between the masses of the electron and the proton. When a new particle with about the predicted mass was discovered in cosmic rays a short time later, this was initially associated with Yukawa's mesotron. However. this particle turned out to be a heavier version of the electron we now call the *muon*, and its ability to penetrate matter to great depths ruled it out as a participant in the strong nuclear force.

Shortly after World War II, a particle now called the *pi-meson* (π), or *pion,* was discovered, also in cosmic rays, whose mass is even closer to the Yukawa prediction. This was the first of the mesons, which now number in the hundreds. This discovery led to a long period of unsuccessful theoretical attempts to quantitatively describe the strong nuclear interaction in terms of meson exchange. Eventually it became clear that the proton, neutron, mesons, and other hadrons were not elementary particles, but were composed of quarks. The strong interaction is now understood in terms of the exchange of gluons between quarks in the theory called quantum chromodynamics.

However, as I have noted, QCD is an unbroken gauge model, which implies that the gluons are massless. Where then does the short range of the strong nuclear interaction come from? QCD provides an answer. The force between two quarks actually increases as they are separated, roughly like a spring force. The intense vacuum energy density that results when the separation distance is of nuclear dimensions results in a breakdown in the vacuum, a kind of lightning bolt that discharges the color field in between and cuts off the range of the force.

On the other hand, the short range of the weak nuclear force, which is much less than the size of a nucleon, is now understood to result from the nonzero masses of the weak bosons. These masses are eighty to ninety times the mass of a proton. Since the weak force, confined to subnuclear dimensions, is the primary source of energy in stars, the Universe would look much different had the symmetries not been broken and the weak bosons had not consequently been provided with mass. Let us now attempt to understand how the symmetries to which we have attributed so much are, in fact, imperfect in the world we live in.

BROKEN SYMMETRIES

It is rather remarkable that the familiar matter of human experience, from rocks and animals to planets and stars, can be reduced to three particles:

u, *d*, and *e*. It is equally remarkable that the interactions of these particles can be reduced to a simple diagram of three lines meeting at a point. Yet the Universe, at least in its current cold state, is not as simple as it might once have been in the early Universe. As shown in fig. 5.1, we find two more generations of quarks and leptons than we seem to need. The higher-generation particles are all highly unstable and not part of familiar matter. Furthermore, we have three types of interactions, plus gravity, when only one might have sufficed.

Yet it appears that these complications are necessary. Otherwise, the Universe might have been too simple for complex systems, such as life, to evolve. For example, without three generations it can be shown that we would not have the large excess of matter over antimatter, by a factor of a billion to one, in the current Universe. Such a universe would have no atoms at all—just pure radiation. So, our Universe may be simple, but only as simple as it can be to still allow for the evolution of life. Or, at least, the point in space and time at which we live provides us with such a subjective point of view that some of the laws we need to describe it necessarily violate point-of-view invariance.

Another complication, besides extra particles, is the mixing of the symmetry of electromagnetism, U(1), with the symmetry of the weak force, SU(2). If these symmetries were perfect, then the three weak bosons would all be massless and the force would have infinite range. Instead, as we have seen, the W and Z bosons are quite massive—comparable to heavy atoms such as silver—and the force confined to subnuclear dimensions.

If gauge symmetry was the star performer of twentieth-century physics, broken symmetry played a significant supporting role, also deserving an Oscar. It had long been assumed that the basic laws of physics obeyed certain symmetry principles, such as the continuous space and time translations and spatial rotation. In addition to these continuous symmetries, the laws seemed to be invariant to certain discrete transformations. In one of these, the handedness of an object is reversed. This operation, which we described above, is called *space reversal* or *parity*, designated by *P*. Space reversal is like looking at a body in a mirror. If the system possesses handedness, or *chirality*, that chirality will be reversed. That is, left turns into right and right into left.

All the laws of classical physics are *P*-invariant. So too were the laws of modern physics—or so it was thought until the 1950s. Then, certain anomalies in the observations of "strange" particles (K-mesons) in cosmic rays led to the suggestion that parity symmetry is violated in weak interactions. This was confirmed by the observation of parity violation in the beta-decay of the radioactive nucleus Co^{60}.

Another symmetry operation that can be also thought of as a kind of mirror reflection involves changing a particle to its antiparticle. This is designated by the operator C. For example, operating with C on an electron gives an antielectron, or positron. C stands for *charge conjugation*, since it reverses the charges of particles. However, it can also act on neutral particles, for example, changing from a neutrino to an antineutrino, so this is a misnomer.

All reactions in which you replace particles by antiparticles are observed to have the same probability as the original reactions. For example, the decay of a negative muon,

$$\mu^- \to e^- + v_\mu + \overline{v}_e$$

occurs at the same rate as the decay of its antiparticle,

$$\mu^+ \to e^+ + \overline{v}_\mu + v_e$$

Note how the final-state particles of the first reaction have been replaced in the second reaction by their antiparticles.

Although weak interactions violate parity symmetry, most are invariant under the combined operation CP. For example, the beta-decay of a neutron

$$n \to p + e^- + \overline{v}_e$$

occurs at the same rate for a given emission direction as the decay of an antineutron

$$\overline{n} \to \overline{p} + e^+ + v_e$$

viewed in a mirror.[2]

In the 1960s it was discovered that certain rare decays of K-mesons were noninvariant under *CP*. The study of the origin of *CP* violation remains a subject of considerable experimental and theoretical effort to this date because of its possible connection with one of the most important symmetry operations, time reversal, designated by the operator *T*.

It can be proved from very general axioms that all of physics is invariant under the combination of operations *CPT*. (In the perspective of this book, we might say that these axioms follow from *CPT* symmetry). It follows that if *CP* is violated, *T* must be violated. Some experimental evidence also exists for direct *T*-violation.

Is this the source of the "arrow of time" of common experience? That hardly seems likely. First of all, the effect is tiny, one part in a thousand, and occurs in just a few very rare processes. Second, what is observed is a small time asymmetry—not time irreversibility. No mechanism has been proposed for how this rare, low probability asymmetry translates into the (almost) irreversible arrow of time of common experience. Third, one can still restore time reversibility in fundamental processes by making sure to perform the *C* and *P* operations as well.

Most physicists prefer to adopt a specific time direction, even when none is called for by theory or experiment. In that case, they must introduce antiparticles. However, a more parsimonious view, which is experimentally indistinguishable from the conventional one, is to interpret antiparticles as particles moving backward in time, that is, opposite the conventional direction.

The alternative pictures are illustrated in fig. 5.4. In (a) we have the "Feynman space-time zigzag," in which an electron goes forward in time, scatters off a photon at time **C**, moves backward in time, scatters again at time **A** and goes forward again.[3] Note that the electron appears simultaneously in three different places at time **B**. In (b), the conventional time-directed view is shown, in which an electron-positron pair is produced at time **A**, with the positron annihilating with the other electron at time **C**. The advantage of view (a) is parsimony, with no need to introduce antiparticles. It also offers an explanation for the indistinguishability of electrons that manifests itself in the Pauli exclusion principle. In (b), the incoming electron is different from the outgoing one. In (a), they are the same electron.

While I find the time-reversible alternative to be a better way to describe physics, to avoid confusion I am adopting the more familiar view in this section.

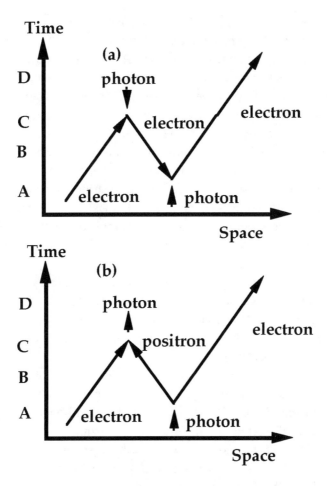

Fig. 5.4. The "Feynman space-time zigzag." In (a), an electron goes forward in time, scatters off a photon, moves backward in time, scatters again, and goes forward. Note that the electron appears simultaneously in three different places at time **B**. In (b), the conventional time-directed view is shown in which an electron-positron pair is produced at time **A**, with the positron annihilating with the other electron at time **C**. The advantage of view (a) is parsimony, with no need to introduce antiparticles. It also offers an explanation for the indistinguishability of electrons.

In any case, observations at the fundamental particle level exhibit some broken symmetry. This should not be unexpected, since the world around us contains violations of many of the symmetries we have been discussing. Earth is round but not a perfect sphere; it is not rotationally symmetric about all axes. Snowflakes have a discrete rotational symmetry in steps of 60 degrees about one axis, but not the continuous symmetry of a disk. Our faces do not possess space reflection symmetry, as a comparison of a photograph of yourself with what you see in the mirror will testify.

So, broken symmetry is a fundamental fact about our Universe. The Universe is not composed of a homogeneous distribution of spherical elementary bodies of zero mass. And that is obviously a good thing, at least from a human perspective. Without this complexity and diversity, the Universe would be a dull place indeed, and, furthermore, we would not be around to be bored by it.

How then did this Universe composed of asymmetric objects come about? What is the mechanism for symmetry breaking? When a direct, causal mechanism can be found, the process is called *dynamical symmetry breaking*. When this is not the case, when no causal mechanism is evident, the process is called *spontaneous symmetry breaking*.

For example, consider a pencil balanced vertically on one end, as shown in fig. 5.5(a). It possesses rotational symmetry about a vertical axis. If someone comes along and pushes the pencil over in a specific, predetermined direction, we have dynamical symmetry breaking. If a (truly) random breeze knocks it over, we have spontaneous symmetry breaking. In either case the pencil now points in a particular direction as in fig. 5.5(b). In the dynamical case, that direction is determined ahead of time. In the spontaneous case, that direction is random. Note that the original symmetry is retained, statistically, in the spontaneous case. If one had an ensemble of pencils and the breezes pushing them over were truly random, then the distribution of directions in the ensemble would remain uniform. Thus, spontaneous symmetry breaking does not destroy the original symmetry of the system; instead it produces a particular case where the symmetry is randomly broken.

Another example of spontaneous symmetry breaking is the ferromagnet (see fig. 5.6). Maxwell's equations are rotationally invariant, yet the lowest energy state of the ferromagnet has a special direction, that of the magnetic field vector. If you start with a spherical ball of iron above a critical temperature, called the *Curie temperature*, and cool it below that point, a phase transition takes you to a lower energy state of broken rotational symmetry. The magnetic domains in the iron act like small bar magnets or *dipoles*. At high temperature they move about randomly so there is no net field. Below the Curie temperature, the mutual interactions

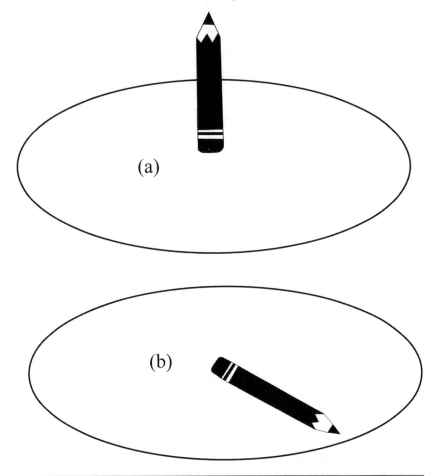

Fig. 5.5. Example of spontaneous symmetry breaking. In (a), a pencil stands on its end and is symmetric about the vertical axis. In (b), a breeze knocks the pencil over and the symmetry is broken, a specific direction in the horizontal plane being selected. If that selection is random, the symmetry is said to be spontaneously broken. Note that the original symmetry is still maintained on a statistical basis.

between dipoles is sufficient to line them up, giving a net field. Again, an ensemble of iron balls would lead to a uniform distribution of magnetic field directions.

The Ferromagnet

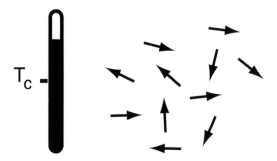

(a) No magnetic field at high temperature.

(b) Magnetic field appears when temperature drops below critical value. Its direction is random.

Fig. 5.6. The ferromagnet is another example of spontaneous symmetry breaking. (a) Above a critical temperature no net magnetic field exists as the magnetic domains inside the iron point randomly and we have rotational symmetry. (b) Below the critical temperature the domains line up, giving a net field and breaking the symmetry.

One of the features of ferromagnets is the generation of magnetic oscillations called *spin waves* when the dipoles are pulled slightly away from equilibrium. When these are quantized, they are described mathematically in terms of particles called *magnons*. In a similar fashion, spontaneous symmetry breaking leads to structures in crystals, and oscillations within those structures result in sound waves, which, when quantized, are described mathematically in terms of particles called *phonons*.

These results from solid-state physics led particle theorists to apply similar notions to the spontaneous symmetry breaking phenomena they were discovering in their arena. There the symmetry breaking was taking place in a more abstract space, but the mathematical tools still applied.

In the standard model, a process called the *Higgs mechanism* generates masses by spontaneous symmetry breaking. The vacuum is postulated to contain a scalar field, which can be thought of as composed of spinless particles whose mass is undetermined in the model. These *Higgs bosons* can be thought of as a Bose condensate filling all of space. None have been observed at this writing, since high energy is required to kick one of them out of the vacuum; but the next generation of experiments at high-energy particle accelerators should either see them or show that they do not exist.

Now, when you put Higgs particles in the equation of motion, that equation is no longer gauge invariant. It can be made gauge invariant by introducing another gauge field. The quanta of this field are particles of spin-1.

Suppose, then, that we have a mixture of spin-0 Higgs bosons and spin-1 gauge bosons. These particles now have mass. This is the mechanism by which the weak bosons W and Z gain mass while maintaining local gauge invariance. They interact with the universal background of Higgs bosons and in the process pick up mass. In simple terms, we can think of a particle gaining an inertial, or an effective mass, by multiple collisions. The Higgs field is analogous to a transparent medium, such as glass, in which photons can be thought of as gaining an effective mass as they scatter from the atoms in the medium. They still travel at the speed of light between collisions but, because of their zigzag trajectory, move through the medium with a net effective speed reduced by the index of refraction of the medium.

The standard model is no doubt just the current step on the way to a fuller theory and has many shortcomings. The point of the present discussion is to show that the basic properties of matter and its interactions are natural consequences of local gauge symmetry, and that more complex properties arise when other symmetries are spontaneously broken while local gauge symmetry is maintained. Let us now take a short look at what might lie beyond, or under, the standard model.

GRAND UNIFICATION

As I have noted, spontaneous symmetry breaking leaves the original symmetry of the fundamental laws intact. The falling over of a pencil does not destroy the rotational symmetry of the laws of gravity. The cooling of an iron ball that leads to a ferromagnet does not destroy the rotational symmetry of the laws of magnetism. Similarly, the symmetry breaking that leads to a mixing of electromagnetism and the weak nuclear force, and generates the masses of the weak gauge bosons, does not destroy the higher symmetry from which these may have arisen.

Usually, when a symmetry is broken, it is not fully broken. What remains are other symmetries more limited than the original but a part of it—subgroups of the original transformation group. For example, a snowflake still has some rotational symmetry but it is about one axis, rather than all axes, and is discrete in 60-degree steps, rather than continuous.

Since the time of the initial successes of the standard model as a theory based on gauge symmetries, theorists have searched for higher symmetries from which the standard model may have arisen. Possibly those symmetries described matter during the early stages of the big bang with the standard model freezing out, like the structure of a snowflake, as the Universe became colder.

Early attempts sought to bring the three forces—electromagnetic, weak, and strong—together as a single force, leaving gravity to be handled later. These were called *grand unification theories* or GUTs. Unfortunately, there are many symmetry groups to choose from that contain $U(1)$, $SU(2)$, and $SU(3)$ as subgroups and only the simplest GUT, *minimal SU(5)*, offered a quantitative prediction that could be tested in experiments to be performed in the immediate future.

Minimal SU(5) predicted that protons would decay at a specific, albeit very low, rate. This prediction was not confirmed, which, while a disappointment, at least demonstrated that this exercise was more than speculation, offering up at least one "good" theory to the altar of falsification. The story is worth reviewing.

Long before the standard model, two conservation principles, which have not been mentioned here so far, had been inferred from observations. The first was *conservation of baryon number*. This was a generalization of conservation of *nucleon number*, which seemed to occur in nuclear reactions. That is, when you counted the total number of protons and neutrons in the atomic nuclei on one side of a reaction, they equaled the total number in the nuclei on the other side.

When particle physicists started observing other hadrons heavier than the proton and the neutron, they found that some reactions did not conserve nucleon number. However, when they assigned a baryon number B to these particles, and used the opposite value for their antiparticles, the total value of B was the same on both sides of all observed reactions. Leptons, such as the electron and the neutrino, mesons, and gauge bosons, were assigned zero baryon number.

The second principle that goes back to the days before the standard model is *conservation of lepton number*. The negatively charged leptons e, μ, and τ and their related neutrinos were assigned lepton number $L = +1$, and their antiparticles $L = -1$. Conservation of lepton number says that the total lepton number is the same on both sides of a reaction. Hadrons and gauge bosons have zero lepton number.

The following reactions exemplify both conservation principles. A negative K-meson can collide with a proton to produce a neutral pi-meson, π^0, and a neutral sigma particle, Σ^0. The sigma and proton each have $B = 1$.

$$K^- + p \to \pi^0 + \Sigma^0$$

The sigma can decay into a lambda, Λ, which also has $B = 1$, along with a positron and an electron neutrino that have $L = -1$ and $L = +1$, respectively:

$$\Sigma^0 \to \Lambda + e^+ + \nu_e$$

The lambda can decay into a neutron with $B = 1$ and a neutral pi-meson:

$$\Lambda \rightarrow n + \pi^0$$

(In all these cases, other decay modes are possible).

Clearly some type of conservation principle is acting to keep the proton stable. If not, it would rapidly decay into lighter particles such as a positron and a photon,

$$p \rightarrow e^+ + \gamma$$

Either baryon number or lepton number conservation would prevent this decay.

In the quark model, baryons are composed of three quarks. Each quark is assumed to have $B = 1/3$ and antiquarks $B = -1/3$, so that the composite baryons have $B = 1$ and antibaryons $B = -1$. An atomic nucleus will have a value of B equal to the number of nucleons it contains.

Baryon and lepton number conservation are embodied in the standard model by the separation of quarks and leptons into distinct sets. This means that quarks do not transform into leptons and vice versa. In most GUTs, this is changed. Quarks and leptons get combined into a single multiplet and interactions can occur that violate B and L conservation.

In particular, minimal SU(5) contains twelve gauge bosons. One of these, called X, has a charge of $-4/3$ times the unit electric charge and allows for a violation of baryon and lepton number and proton decay, as illustrated in fig. 5.7. The theory predicted a decay rate for the proton by this reaction of one every $10^{30\pm1}$ years. A number of experiments were launched to search for proton decay but so far have come up negative. The current experimental limit on the particular decay mode shown in fig. 5.7 is greater than 10^{33} years. Other decay modes have also been searched for with the limits greater than 10^{31} years.

So minimal SU(5) is falsified as a possible GUT. But, as mentioned, there are still many other possibilities. Indeed, the search will probably continue since there is in fact strong evidence for baryon nonconservation. The visible Universe contains a billion times as many protons as antiprotons. Baryon number conservation would demand they be equal. Of course, it could have just happened that way; the Universe might have

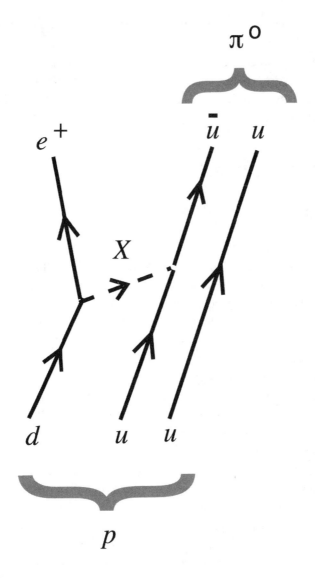

Fig. 5.7. Proton decay into a positron and a neutral pion, as predicted by grand unified theories (GUTs), but not yet observed in experiments. The *X* is a gauge boson of the theory, not part of the standard model, which allows for the violation of baryon number and lepton number conservation.

been born with an excess of baryons. More likely, equal numbers of baryons and antibaryons existed in the early stages of the Universe, when the symmetries were higher, and then, as suggested years ago by famed Russian physicist (and political figure) Andrei Sakharov, baryon nonconserving processes produced an imbalance. A similar explanation would hold for the excess of leptons over antileptons.

SUPERSYMMETRY

Supersymmetry (referred to affectionately as SUSY) is the notion that the laws of physics are the same for fermions and bosons. It implies that every fermion has associated with it a SUSY boson and, likewise, every boson has associated with it a SUSY fermion. The super partners are called *sparticles*. Thus the spin-1 photon has a spin-1/2 partner called the *photino*. The spin-1/2 electron has a spin-0 partner, the *selectron*. The spin-1/2 quarks have spinless *squark* partners, the spin-1 gluon has the spin-1/2 *gluino*, the spin-1 W-bosons have the spin-1/2 *wino*, and so on.

Obviously supersymmetry, if it is valid at all, is a broken symmetry at the low energies of our experience and current experimentation, since sparticles have not been observed at this writing. Some rules of physics, such as the Pauli exclusion principle, are dramatically different for fermions and bosons. Furthermore, the fact that no sparticles have yet been observed implies their masses will be greater than several hundred GeV (1 GeV = 10^9 electron-volts), certainly not equal in mass to their partners, as unbroken supersymmetry would imply. If the current observable Universe were supersymmetric, it would be a far different place.

Sparticles will be searched for intensely in the next generation of particle accelerators. Supersymmetry has many theoretical features that make it extremely attractive as an underlying symmetry of nature. It neatly provides answers to several questions that are unanswered in the standard model.[4]

SUSY may also provide an answer to a current cosmological puzzle. Only 3.5 percent of the mass-energy of the Universe is composed of familiar matter, that is, electrons and nuclei (only 0.5 percent is visible). This conclusion is drawn from calculations of the synthesis of deuterium, helium, and lithium in the early Universe. These calculations are regarded as very reliable since they are based on well-established low-energy nuclear physics.

SUSY is a common ingredient in various current attempts to produce a quantum theory of gravity. It is fundamental to string theory, which seeks to unify all the forces of nature. And it provides possible candidates for that major portion of the dark matter of the Universe, since calculations on big-bang nucleosynthesis show it cannot be "baryonic." These dark matter candidates are called *WIMPS*—weakly interacting massive particles. Several SUSY particles are candidates for these particles.

A supersymmetric version of SU(5), which I will call SU(5)SY, has provided one result that has encouraged many theorists to believe that they are on the right track toward unifying the forces of nature by means of supersymmetry. Like minimal SU(5) discussed above, SU(5)SY is a minimal supersymmetric model having the fewest unknown parameters in this class of models.

In grand unified theories, the parameters that measure the strengths of the various forces vary with energy. That is, they are said to "run." SU(5)SY predicts that the inverse of the dimensionless force strengths will vary linearly with the logarithm of energy. Fig 5.8 shows a plot of the three inverse strengths, where the measured values of the constants at 100 GeV, the highest current energy, are used as fixed points. The three lines are extrapolated independently over 14 orders of magnitude in energy using the slopes given by the model. Remarkably, the lines meet at a "unification energy" of 3×10^{16} GeV. Such a convergence is not observed for minimal SU(5) without SUSY.

Of course, two nonparallel lines will meet someplace. However, the simultaneous crossing of three lines after 14 orders of magnitude of extrapolation seems highly unlikely to be accidental. This suggests that above 3×10^{16} GeV, supersymmetry is in effect with the forces unified. Below that energy the symmetry breaks and the electroweak and strong forces differentiate. Unfortunately, the unification energy is far from anything that is likely to be achieved in experiments for the near future, so any effects of this unification will have to be inferred indirectly at lower energies.

It is not my purpose here to indulge in speculations about the physics beyond the standard model, or even to discuss all the details of that highly successful theory. This limited discussion is meant to illustrate that symmetries play a basic role in all modern theories in particle physics and are likely to do so in the future. Without knowing the exact nature of the underlying symmetries out of which the current standard model crystallized, we can safely assume that some high level of symmetry describes an underlying reality that is far simpler than what we observe in today's frozen Universe.

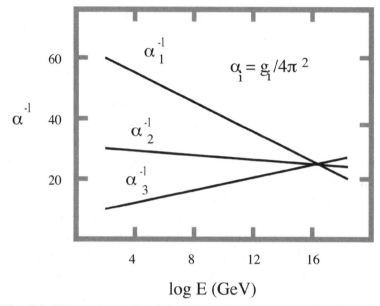

$$\log E \text{ (GeV)}$$

Fig. 5.8. The reciprocals of the dimensionless strengths of the interactions in the standard model are predicted to vary linearly with log energy in the supersymmetric grand unified theory. The measured values at 100 GeV are used and the lines independently cross at a unification energy of 3×10^{16} GeV.

THE TABLET OF PHYSICAL LAWS

We are now in a position to inscribe a tablet of the basic laws of physics that can be placed on the lawn in front of government buildings instead of the Ten Commandments that some would like to see there.

What has been attempted here is to show that the laws of physics do not follow from very unique or very surprising physical properties of the Universe. Rather, they arise from the very simple notion that whatever mathematics you write down to describe measurements, your equations cannot depend on the origin or direction of the coordinate systems you define in the space of those measurements or the space of the functions used to describe those laws. That is, they cannot reflect any privileged point of view. Except for the complexities that result from spontaneously broken symmetries, the laws of physics may be the way they are because they cannot be any other way. Or, at, least they may have come about the simplest way possible. Table 5.1 summarizes these conclusions.

Table 5.1. The Laws and Other Basic Ideas of Physics and Their Origins

Law/Idea of Physics	Origin
Conservation of momentum	Space translation symmetry
Conservation of angular momentum	Space rotation symmetry
Conservation of energy (first law of thermodynamics)	Time translation symmetry
Newton's 1st law of motion	Conservation of momentum (space translation symmetry)
Newton's 2nd law of motion	Definition of force
Newton's 3rd law of motion	Conservation of momentum (space translation symmetry)
Laws of rotational motion	Space rotation symmetry
Second law of thermodynamics	Statistical definition of the arrow of time
Special relativity	Space-time rotation symmetry
Invariance of speed of light	Space-time rotation symmetry
General relativity	Principle of covariance (space-time symmetries)
Quantum time evolution (time-dependent Schrödinger equation)	Global gauge invariance
Quantum operator differential forms	Global gauge invariance
Quantum operator commutation rules	Global gauge invariance
Quantization of action	Global gauge invariance
Quantization rules for angular momenta	Global gauge invariance
Maxwell's equations of electromagnetism	Local gauge invariance under $U(1)$

Quantum Lagrangians for particles in presence of electromagnetic field	Local gauge invariance under U(1)
Conservation of electric charge	Global gauge invariance under U(1)
Masslessness of photon	Local gauge invariance under U(1)
Conservation of weak isospin	Global gauge invariance under SU(2)
Electroweak Lagrangian	Mixing of U(1) and SU(2) local gauge symmetries (spontaneous symmetry breaking)
Conservation of color charge	Global gauge invariance under SU(3)
Strong interaction Lagrangian	Local gauge invariance under SU(3)
Masslessness of gluon	Local gauge invariance under SU(3)
Structure of the vacuum (Higgs particles)	Spontaneous symmetry breaking
Doublet structure of quarks and leptons	Conservation of weak isospin (global gauge invariance under SU(2))
Masses of particles	Higgs mechanism (spontaneous symmetry breaking)

NOTES

1. For an excellent undergraduate textbook on elementary particle physics, see David Griffiths, *Introduction to Elementary Physics* (New York: Wiley, 1987).

2. By this I simply mean that one reverses the chiralities, as in a mirror image.

3. Richard Feynman, "Space-Time Approach to Non-relativistic Quantum Mechanics," *Reviews of Modern Physics* 20 (1948): 367–87.

4. Gordon Kane, *Supersymmetry: Unveiling the Ultimate Laws of Nature* (Cambridge, MA: Perseus), pp. 56–61.

CHAPTER SIX

PLAYING DICE

PARTICLES AND THERMODYNAMICS

While the notion that matter is composed of localized objects that move around in an otherwise empty void is an ancient one, going back to Democritus (d. 370 BCE), it was not fully validated (as a useful model) until the twentieth century. Early in the previous century, John Dalton (d. 1844) proposed an atomic model that implied the existence of primary chemical "elements." However, the particulate nature of these elements was not immediately clear. As we now know, they are far too small to have been seen individually with the technology of that day. Many scientists at that time followed the lead of Ernst Mach, who declared that he would not believe atoms existed until he saw one. Unfortunately, Mach did not live to see the pictures of atoms that are today routinely obtained with scanning tunneling microscopes (STM).

The smallness of atoms also means that huge numbers of them are present in even the tiniest amount of matter. Thus, the statistical fluctuations you expect with a particulate scheme are often too tiny to be observed. For example, suppose we have a gas with only one hundred particles. The instantaneous pressure of this gas will fluctuate by an easily observed 10 percent about its average. For typical macroscopic systems, however, the number of particles is on the order of 10^{24}. In that case, the pressure fluctuation is only one part in a trillion and not likely to be noticed. The practical result is that, for most purposes, macroscopic thermodynamic quantities, such as the average pressure and mass density, can be treated as if they are exact.

To the nineteenth-century observer, then, bulk matter appeared continuous and operationally defined variables, such as pressure, density, and temperature, could be used to describe a body as a whole. Independent of any assumption of atomicity, observations led to the development of theoretical principles that successfully described the behavior of practical macroscopic systems.

In the meantime, Newtonian mechanics was extended to the description of rigid bodies, which represents a good approximation to most everyday solids when temperature and pressure variations are small. Under those conditions, the motion of a solid body of arbitrary shape can be described as the motion of a particle of the same mass located at the center of mass of the body, along with any rotational motion about that point in the center of mass reference frame. Liquids and gases can be modeled in terms of fluid elements that also move according to Newtonian principles. In today's elementary physics courses, the methods of continuum calculus are used to derive the equations that describe the motion of rigid bodies and fluids directly from particle mechanics. This is not done at the molecular level, but rather with small volume elements containing many molecules. Those volume elements move around just like particles.

Although Benjamin Thompson, Count Rumford (d. 1814) had demonstrated the mechanical equivalent of heat and energy in the eighteenth century, heat appeared to be a special form of energy, different from the familiar mechanical energy that characterized the motion of solids and incompressible fluids. The science of thermodynamics was developed to describe heat phenomena, with the very practical application to the understanding of heat engines and refrigerators that were beginning to be developed by industrial society.

Classical thermodynamics is based on three principles. It starts with what is often called the *zeroth law of thermodynamics*, but it is really just a definition. A body is said to be in thermal equilibrium when it has the same constant temperature throughout. Temperature is operationally defined as what you read on a constant-volume gas thermometer.

The first law of thermodynamics describes the special form that conservation of energy takes when dealing with heat phenomena. The heat added to a system is equal to the change in the internal energy of the system, plus the work done by the system. Two systems in contact can exchange energy back and forth in the form of heat or work. Note the convention that work is positive when the work is done by the system. It will be negative when the work is done on the system. Normally the pressure is positive (but not always—negative pressures can occur), so work is done by a system, such as a gas when it expands against some other medium.

In general, you can have both heat and work contributing to changes in internal energy. A temperature differential between two bodies in contact will result in a flow of heat from one to the other. One body will lose some internal energy and the other will gain some, so heat is seen as an energy flow. The first law "allows" this flow to happen in either direction. However, observations show something different. When no work is done on or by either body, heat always flows from a higher temperature to a lower temperature. If this were not the case, we could install refrigerators and air conditioners that need not be plugged into the power receptacle on the wall.

Similarly, the first law does not "forbid" the energy in our surrounding environment from being used to run a heat engine. But our experience tells us that we have to regularly pull into a filling station to fuel our cars, that perpetual-motion machines do not exist.

The second law of thermodynamics was introduced in the nineteenth century in order to describe these empirical facts. The simplest form of the second law, proposed by Rudolf Clausius (d. 1888), says that in the absence of work, heat always flows from a higher temperature to a lower temperature. That is, one cannot build a perfect refrigerator or air conditioner, one that can cool a body below the temperature of its environment without doing work on the system.

The Clausius version of the second law can be shown to be equivalent to an alternate version proposed by William Thomson, Lord Kelvin.[1] Kelvin asserted that a heat engine cannot convert all of the energy from an external energy source into useful work. That is, one cannot build a perfect heat engine or what amounts to a perpetual-motion machine. This form of the second law follows from the Clausius form, since if heat could flow from a lower to a higher temperature without doing work, you could take the exhaust heat of a heat engine and feed it back to the higher-temperature source. Similarly, Clausius's form follows from Kelvin's, since if the latter were not true, you could build a perfect refrigerator by using a perfect heat engine to provide the work.

Sadi Carnot (d. 1832) provided a mathematically precise form of the second law by stating it in terms of a quantity called the *entropy* that can loosely be identified with the disorder of the system. In terms of entropy, the second law then says that the entropy of an isolated system must remain constant or increase with time. It can never decrease.

STATISTICAL MECHANICS

In the late nineteenth century, James Clerk Maxwell, Ludwig Boltzmann, and Josiah Willard Gibbs found that the laws of classical equilibrium thermodynamics could be derived from Newtonian particle mechanics on the assumptions that matter was composed of atoms and that all the allowed states of a system are equally likely.[2]

The pressure on the walls of a chamber containing a fluid was interpreted as the average effect of the molecules of the fluid hitting the walls and transferring momentum to those walls. The absolute temperature was identified with the average kinetic energy of the molecules. Thermal equilibrium, as defined by the zeroth law, corresponded to the situation where the molecules of a body have moved around and collided with one another for such a sufficiently long time that any group of them in one region of the body has the same average kinetic energy as a group in some other region. This is clearly a special circumstance; any given system of molecules will not instantaneously reach thermal equilibrium when its conditions are abruptly changed, such as by a rapid expansion of the volume or a sudden injection of heat. The atomic model makes it possible to study nonequilibrium systems in a way not available in classical thermodynamics.

The first law is also easily reconciled in the atomic model. The internal energy of a body is the sum of the kinetic and potential energies of the molecules of the body, including any rotational and vibrational energy of the molecules themselves.

The energy of a system of molecules can change in several ways. Work can be done on the system, as in the compression of a gas. Work can be done by the system, as in the expansion of a gas. These are normal mechanical processes. Compressing the gas, for example, results in a smaller volume and a more rapid rate of molecular collisions with the wall as the molecules have less distance to cover as they bounce around. Expanding the gas has the opposite effect.

In classical thermodynamics, heat is treated as a different form of energy that can also be added to or subtracted from the system without mechanical motion. In the atomic picture, however, heat is simply mechanical energy carried from one body to another. This can happen in a number of ways. When two bodies are in contact, the molecules at the interface collide and exchange energy. This form of heat transfer is called conduction. When two bodies are separated but some medium, such as air, connects them, heat can flow by convection as air molecules receive energy by collisions from the hotter body and move across the space, transmitting energy by collisions to the colder body. Heat can also be transmitted by radiation, such as from the Sun. Here photons move across the space between and transfer their kinetic energy to the absorbing body.

The second law also follows from the basic Newtonian principles by which atoms interact in macroscopic bodies. Statistical mechanics can be use to derive Boltzmann's *H-theorem*, which shows that a molecular system originating arbitrarily far away from equilibrium will approach equilibrium and that the entropy will be maximal at that point.

Indeed, all of equilibrium thermodynamics is derivable from particle mechanics, with the added assumption that all the allowed states of a system are equally likely. This includes not only classical thermodynamics but quantum thermodynamics as well, when we understand that we are dealing with systems in equilibrium. Nonequilibrium systems, both classical and quantum, are much more difficult to deal with, although much progress has been made. Much of that progress has been made possible by the use of high-speed computers that can simulate the complex motions of particles that otherwise cannot be treated by familiar mathematical methods. Those details are beyond the scope of this book. However, it is to be remarked that none of the discoveries that have been made in the study of complex, nonequilibrium systems requires any additional principles of physics than those discussed in this book.

TIME SYMMETRY AND CAUSALITY

The second law of thermodynamics says that the entropy of an isolated system must stay the same or increase with time. But how do we define the direction of time? All the equations of physics, except for the second law, work in either time direction. No preference for one time direction over the opposite has been found in any fundamental process, although, to be strictly accurate, in a few very rare cases you must exchange a particle with its antiparticle and reverse its handedness to maintain perfect time symmetry.

As Boltzmann noted, increasing entropy can in fact be regarded as defining what Sir Arthur Eddington dubbed the *arrow of time*. Thus we have another "law" of physics that turns out to be nothing more than a definition. However, as we saw previously, the second law is a statistical statement that is meaningful only when the number of particles is large. This implies that the arrow of time only applies in that context. In systems with few particles, the entropy will fluctuate substantially from its equilibrium value and an arrow of time becomes impossible to define.

A slight time asymmetry in reaction rates on the order of 0.1 percent is observed in the weak interactions of elementary particles. This does not account for the macroscopic arrow of time. Time asymmetry does not forbid reverse processes from occurring; the reaction rate in one time direction is simply different than in the other. Even this slight asymmetry goes away if, as mentioned above, you also change all the particles to antiparticles and take the mirror-image event: that is, you perform the combined operation CPT described in chapter 5 (see fig. 5.3).

In a previous book, *Timeless Reality: Symmetry, Simplicity, and Multiple Universes*,[3] I discussed in detail the implications of the apparent time-reflection symmetry of the fundamental laws of physics. In particular, I showed that many, if not all, of the so-called paradoxes of quantum mechanics can be understood as a consequence of forcing the familiar time direction of ordinary experience on the description of quantum events. Quantum time reversibility does not mean that humans can travel back in time or that time-travel paradoxes do not exist for macroscopic systems. But what it does mean is that one should not assume a direction of time when one is talking about the fundamental processes of nature. The common notion that cause always precedes effect must be modified, so that cause and effect are interchangeable. What we call the effect can just as well be the cause, and the beginning can just as well be the end.

Indeed, the whole notion of cause and effect has been questioned by philosophers since David Hume (d. 1740) in the eighteenth century.[4] More recently, philosopher Bede Rundle has remarked, "Causation is thought of as a relic of a primitive view of viewing nature—perhaps one centered in human action—which is to be replaced by the notion of functional dependence." He describes what we conventionally think of as a causal sequence of events as a "regularity of succession," noting that "if [that] is all there is to causation, why not allow that a cause *should* act in reverse."[5]

RANDOMNESS

We have seen that a system of particles is in equilibrium when its entropy is maximal. In statistical mechanics, classical equilibrium thermodynamics is derived from particle mechanics and the assumption that all the allowed states of a system are equally likely. Physical configurations and conservation principles set a boundary in which the allowed states of the system are confined. For example, particles may be confined to an insulated container so that no particle can be outside the box or have more than a maximum momentum. However, when that system is in equilibrium, each state within that region is equally likely. Which state is actually occupied is then a matter of chance.

We can visualize this in terms of phase space, which was introduced in chapter 4. There we saw that the state of a classical system of N particles, with no internal degrees of freedom, such as spin or vibration, is a point in an abstract space of $6N$ dimensions, where $3N$ dimensions give the spatial coordinates of the particle and the other $3N$ dimensions give all the particle-momentum components. In fig. 6.1. we show a two-dimensional plot of the x-coordinate of one particle and its momentum in the x-direction, p_x. The state of the particle is confined to a point within the box, which sets the limits on x and p_x. If nothing otherwise selects a preferred state, every point within the box is equally likely.

Let us look further at the role of chance in physics. In the Newtonian picture, the state that a system of particles in equilibrium attains is, in principle, predictable from the initial conditions and an application of the deterministic laws of physics. In the case of macroscopic systems where we are dealing with 10^{24} or so particles, it becomes a practical impossibility to calculate the motion of each particle. Still, we have no reason to require all states to be equally likely in a completely deterministic uni-

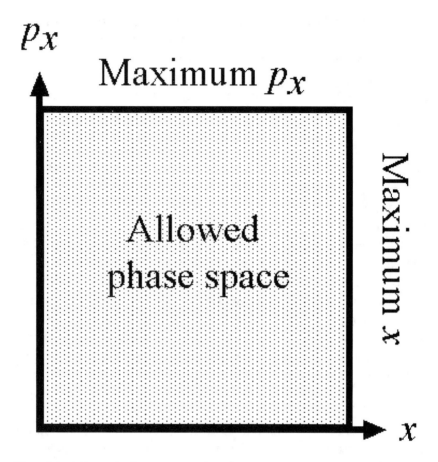

Fig. 6.1. Allowed phase space for particles confined to an insulated box. Just one dimension of one particle is shown. The maximum x is determined by the size of the box, and the maximum p_x by the total energy of the particles.

verse. We can expect every state to be sampled over a long period of time, but the density of states would not necessarily be uniform.

With quantum mechanics, however, we have reason to expect that systems of many particles will eventually fill all phase space with uniform density. The uncertainty principle places a limitation on how precise particle paths can be determined, implying a certain amount of inherent randomness. At least, this is what is implied in most interpretations of quantum mechanics. The one exception is the nonlocal, deterministic interpretation in terms of hidden variables introduced by David Bohm.[6] Here a particle's motion is determined by its instantaneous (and superluminal) interaction with all the other particles in the Universe. The apparent randomness involved in that motion is then like that in classical deterministic statistical mechanics, the net effect of a huge number of interactions. However, as in the classical case, no good reason exists for the observed smooth phase-space density of equilibrium systems.

While Bohmian quantum mechanics has some supporters, the majority of physicists and philosophers are unsympathetic because of its implicit violation of Einstein's principle that no signals can move faster than the speed of light. Until such a violation is observed empirically, or some evidence is obtained for subquantum forces, Bohmian quantum mechanics remains a nonparsimonious alternative to indeterministic quantum mechanics.

Random numbers are used in many applications, such as computer Monte Carlo simulations and cryptography. However, these generally use computer-generated sequences of numbers that are produced by the execution of mathematical algorithms. The numbers produced are not regarded as "truly random" and are called *pseudorandom numbers.*

Then what is a "truly random" number? A simple definition is a number that cannot be generated by any algorithm that is shorter than the number itself. An algorithm, or mathematical procedure, can be expressed as a sequence of binary bits, as can any number. The *Kolmogorov complexity* of a number is defined as the minimum number of bits needed for an algorithm to generate the number. Such a number can be shown, in the infinite-string limit, to pass all the statistical tests for randomness.

The problem is: How can we obtain a number that is algorithmically incompressible? The solution is: We take it from nature. For example, we can construct a random-number generator with a radioactive source. As best as we can tell from experiment, the probability for a radioactive nucleus decaying in given time interval is independent of the temporal location of that time interval. This is what leads to the exponential-decay "law" that is observed to great precision. A simple random-number generator is then a Geiger counter whose output rate is digitized and the logarithm taken of the resulting number.

CHANCE

The role that randomness plays in the world is called *chance*. Chance is the antithesis of *necessity*. The ancient notion that the Universe is ruled completely by laws, natural or divine, leaves no room for chance. In a deterministic universe, classical or Bohmian, every event is predetermined by prior (or past) events. In an indeterministic universe, such as one described by conventional quantum mechanics, they are not. Chance is the name we give to the process by which events happen in a way not fully determined by prior (or, in a time-symmetric universe, later) events. Evidence for the role of chance is found in the behavior of many body systems, as discussed above, in quantum phenomena, such as radioactive decay and atomic transitions, and in Darwinian evolution. In fact, statistical thinking developed in the social sciences before the recognition of its role in physics and biology.

Deterministic physics is seen as a special case of the physics of chance and necessity. In the case of pure necessity, that is, absolute determinism, the probability for an event is either zero or one. The evidence strongly indicates that this is not the universe we live in but that most probabilities lie between the extremes. Chance plays a big role—a much bigger one than most people think.

In fact, as we will see in the next chapter, most of the Universe is composed of particles that move around randomly. For example, for every atom of familiar matter in the Universe, a billion photons are found in the cosmic microwave background. These photons form a gas that is in thermal equilibrium at 2.7 degrees Kelvin to one part in a hundred thousand. Indeed, the visible matter in stars and galaxies, which seems to testify to an orderly cosmos, constitutes only half of 1 percent of the mass of the Universe.

As we have already seen in great detail, the "necessity" part of the process does not conform to the widely held notion of a universe bound by externally enforced rules of behavior. Rather, the necessity is one forced on physicists when they mathematically formalize the patterns they see in their quantitative measurements. We have seen that these formulations are chosen so as not to depend on the point of view of any observer, that is, they possess certain symmetries. As far as we currently know, there are no indications that the matter of the Universe is governed by any external, metaphysical agency that forces it to behave in a specific way. Even those symmetries that are broken at the low energies of common experience and current scientific investigation are maintained statistically. The apparent order we find, as described by the global laws of physics, is forced upon us by the very absence of governance.

NOTES

1. My convention is to only give the year of death when it occurred prior to the twentieth century and only the first time the name appears in the book.

2. For an excellent undergraduate-level exposition, see F. Reif, *Fundamentals of Statistical and Thermal Physics* (New York: McGraw-Hill, 1965). The classic graduate textbook is Richard C. Tolman, *The Principles of Statistical Mechanics* (London: Lowe & Brydone, 1938). Later printings are available from Oxford University Press.

3. Victor J. Stenger, *Timeless Reality: Symmetry, Simplicity, and Multiple Universes* (Amherst, NY: Prometheus Books, 2000).

4. David Hume, *An Enquiry concerning Human Understanding* (1748), ed. Tom L. Beauchamp (Oxford: Oxford University Press, 1999).

5. Bede Rundle, *Why There Is Something Rather than Nothing* (Oxford: Clarendon, 2004), pp. 67–68.

6. D. Bohm and B. J. Hiley, *The Undivided Universe: An Ontological Interpretation of Quantum Mechanics* (London: Routledge, 1993).

CHAPTER SEVEN

AFTER THE BANG

THE EXPANDING UNIVERSE

So far, we have limited our discussion to the familiar models of physics and how the so-called laws that are contained in these models arise naturally from the principles of point-of-view invariance and maximum simplicity. One of these models is general relativity, which has its most important applications in astronomy and cosmology. In this chapter, we will see how a combination of general relativity and more recent models of cosmology and particle physics now provide a consistent, understandable picture of the cosmos that agrees with all observations. Furthermore, these well-established models can be extrapolated to provide scenarios for the early Universe that plausibly account for the origin and structure of matter in terms of its evolution from an initial highly symmetric, homogeneous, and totally disorganized state. This will support the view presented in the preceding pages that the laws of physics arise as physicists describe observations in terms of symmetries and the accidental breaking of these symmetries in order to maintain point-of-view invariance. Let us begin with the data.[1]

In the early twentieth century, astronomer Edwin Hubble discovered that the Universe is expanding as if its constituents were part of a gigantic explosion that astronomer Fred Hoyle later derisively coined the "big bang." In 1980 the big-bang model was modified to include an initial, very short period of exponential expansion called *inflation*.[2] More recent observations have solidly supported this picture.

Hubble, and many astronomers afterward, measured the *redshifts*—fractional changes in wavelength of light emitted from galaxies, which for speeds low compared to light is proportional to the recessional speed of a galaxy. One way to produce a redshift is by the *Doppler effect*, in which the frequency of light is decreased (redshifted) when the source is moving away, and increased (blueshifted) when it is moving toward an observer. Most galaxies are moving away, although our nearest neighbor, Andromeda, is actually moving toward us.

Hubble found that, on average, the speed at which a galaxy recedes from us is proportional to its distance, as one would expect from an explosion in which the faster moving fragments go farther than slower ones. The proportionality factor is called the *Hubble constant*. Its inverse, converted to a unit of time, is about 13.8 billion years, which gives the age of the Universe if we assume the Hubble constant has not changed substantially since the start of the big bang. Actually, as we will see, it is currently changing slightly with time, although a 14-billion-year age estimate is adequate for our purposes. Furthermore, the value of the Hubble constant may have changed very rapidly in the early Universe. Thus a better term is *Hubble parameter.*

Many years of intense effort were required to obtain accurate, consistent data on the current value of the Hubble parameter. Many difficult instrumental problems had to be solved and independent techniques had to be developed to measure cosmological distances. Only recently have estimates converged. This complex story is related in many books on astronomy and need not be repeated here.

Since Hubble's initial observations, data supporting the big bang have continued to accumulate. The few remaining big-bang skeptics are slowly dying out. The evidence has overwhelmed the skepticism still occasionally reported in the popular media. Today, no one involved in serious cosmological research has any doubt that the big bang happened.

GENERAL RELATIVITY AND COSMOLOGY

Einstein's general theory of relativity had profound cosmological implications. When applied for speeds low compared to the speed of light, his equations reproduced those of Newtonian gravity. Einstein additionally found that light is also attracted gravitationally to massive bodies. His famous prediction that light would be bent in passing by the Sun was confirmed by observation. In the intervening years, Einstein's theory has passed many stringent empirical tests.

The most familiar ingredients of the Universe are matter and radiation. Normal matter, such as the stuff in stars and planets, is composed of bodies moving at speeds low compared to the speed of light. It is "nonrelativistic" matter. In cosmology, *radiation* refers to material bodies that move at speeds at or near the speed of light. It is relativistic matter. The pressure of nonrelativistic matter is negligible compared to its density (these have the same units when $c = 1$). The pressure of radiation, on the other hand, equals one-third its density.

In general relativistic cosmology, the expansion of an assumed homogeneous and isotropic universe is described by a *scale factor* that multiplies the distances between bodies as a function of time. General relativity says that the total mass-energy of nonrelativistic matter inside an expanding sphere is a constant, and its mass-energy density then falls off as the cube of the scale factor. The total energy of radiation, however, is not constant inside such a sphere and its density falls off as the fourth power of the scale factor.

THE CRITICAL DENSITY OF THE UNIVERSE

In Einstein's formulation of general relativity, the geometry of space-time is *non-Euclidean*. That is, the sum of the interior angles of a triangle formed, say, by three beams of light, does not necessarily add up to 180 degrees.

However, a *critical density* can be defined as the average density of the Universe when space is "flat," that is, the geometry of space is, on average, Euclidean. With the current estimate of the Hubble parameter, the critical mass density is 9.5×10^{-30} g/cm^3 and the critical energy density (using $E = mc^2$) is 5.3×10^3 electron-volts/cm^3.

Astronomical observations have long indicated that the density of *visible* matter is far less than critical. In the meantime, evidence for a nonluminous component pervading space, called *dark matter*, has gradually accumulated. For a while it was thought that the dark matter might provide the necessary mass to achieve criticality. However, this is not the case.

In the last decade, observational cosmology has become a science of unprecedented precision, especially with the detailed mapping of the cosmic microwave background, which is extremely sensitive to cosmological parameters. These measurements strongly demonstrate that the average mass density of the Universe is very near the critical value. At the same time, other improved observations have shown that the dark matter is insufficient to reach this level. At one point, around the turn of the millennium, it seemed that a major disagreement existed among observations, with the cosmic microwave background indicating critical density but astronomers unable to find sufficient matter to reach that level.

The inflationary model of the early Universe requires that the average mass density be critical to high precision. According to this model, during the first tiny fraction of a second, the Universe underwent an enormous exponential expansion. Inflation thus implies that the Universe is, on average, geometrically flat to a very high degree. This is not hard to understand: If we imagine inflating a balloon to huge dimensions, a small patch on its surface will be flat to a good approximation. Of course, the surface of a three-dimensional balloon is two-dimensional. Our Universe is like a tiny three-dimensional patch on a four-dimensional balloon, in which the additional dimension is time. Inflation would almost certainly be falsified if any major deviation from critical density was found. Hence, with the dark matter proving insufficient, inflation was in deep trouble until an amazing, unexpected discovery bailed it out.

THE UNIVERSE IS FALLING UP!

Until recently, measurements of the speeds at which galaxies recede from us, plotted against their distances, have fit a straight line of constant slope, although individual points showed a lot of scatter. That slope is just the Hubble parameter. Because of the finite speed of light, we are seeing the most distant galaxies as they were long ago, so a straight-line plot would indicate that the Hubble parameter has remained constant over billions of years. Astronomers anticipated, however, that they would eventually see a slight upward curving in the line indicating higher recessional speeds in the past. A slowing down with time was predicted as the mutual gravitational attraction of galaxies acted to brake their relative motions. Observers kept looking and have now finally seen the line curve, but it is sloped the "wrong" way—downward. In 1998 two independent groups

Table 7.1. The Mass-Energy Budget of the Universe	
Radiation	0.005%
Ordinary visible matter	0.5%
Ordinary nonluminous matter	3.5%
Exotic dark matter	26%
Even more exotic dark energy	70%

using supernova explosions in galaxies as standard candles made the unanticipated discovery that the expansion of the Universe is not slowing down; rather, it is speeding up![3] Galaxies are now moving away from each other faster than they did in the past. This discovery has been further independently verified, most importantly by the Hubble Space Telescope.

From the detailed analyses of the wealth of recent cosmological data, the positive mass-energy (rest plus kinetic) density of the Universe is shared among its various components as indicated in table 7.1 (these numbers can be expected to change slightly as observations improve).

The agent responsible for acceleration is called *dark energy*. As we see from the above table, this represents the largest component of the Universe. What could this dark energy be?[4]

GRAVITATIONAL REPULSION

When Einstein was developing general relativity, the common understanding was that the Universe is a firmament of fixed stars. But, as had already been recognized by Newton, such a firmament is gravitationally unstable. In order to achieve a stable firmament, a repulsive force was needed to balance the attractive force of gravity. "Antigravity," while a common topic in science fiction, had never been observed.

General relativity, however, allows for a gravitational repulsion if the pressure of the gravitating medium is sufficiently negative—less than minus one-third its density. Despite the fact that negative pressures occur in some material media, Einstein regarded the prospect of negative pressure in cosmic matter as unphysical. The stars, planets, and dust in the cosmos have positive pressure, as does relativistic matter, such as the photons in the cosmic microwave background radiation. Even the dark matter, whatever it is, has positive pressure. A new form of cosmic matter would seem to be needed to produce a gravitational repulsion.

Einstein discovered a means to achieve the repulsion needed to stabilize the Universe without introducing exotic media. He noted that his equation for space-time curvature allowed for the inclusion of a constant term called the *cosmological constant*. When this term is positive, the pressure equals exactly minus one-third the density and gravity is repulsive.

The cosmological term does not correspond to any physical medium but rather the curvature of space-time itself. Empty space can still have curvature. This solution of general relativity is called the *de Sitter universe*.

When Hubble discovered that the Universe was not a stable firmament after all, but an expanding system, Einstein withdrew the cosmological term, calling it his "greatest blunder." It was hardly a blunder. The cosmological constant is allowed by general relativity and no currently well-established principle of physics rules it out. In the 1980s the possibility of gravitational repulsion once again surfaced in cosmology with the proposal of a rapid acceleration, called *inflation*, during the first tiny fraction of a second in the early Universe and the discovery that the current Universe is undergoing a much slower accelerating expansion.

A vacuum contains no matter or radiation. However, the energy density of the vacuum itself is not necessarily zero, in which case the pressure of the vacuum is negative and equal to minus one-third of the density. In general relativity, an energy density is equivalent to the curvature of space-time. Meanwhile, quantum mechanics implies that the vacuum contains nonzero energy in the form of pairs of particles that flit in and out of existence on short time scales, which can, in principle, account for the nonzero energy density of the vacuum (see chapter 8). So, a meeting of the minds between quantum mechanics and general relativity is implied.

COSMIC ACCELERATION VIA THE COSMOLOGICAL CONSTANT

Let us look at the acceleration that occurs by means of the cosmological constant. In a universe empty of matter and radiation but with negative pressure equal to minus the density, the scale factor of this universe is in general the sum of two exponentials, one increasing and one decreasing with time, where zero time is defined as the initiation of the big bang. In most presentations, the decreasing exponential is neglected since it rapidly goes to zero as time increases. However, the roles of the two exponentials are reversed at negative times, that is, times *before* the initiation of the big bang. There is no fundamental reason why these equations should not apply equally well for negative time as for positive time.

More generally, in a universe including matter and radiation, a net acceleration will occur when the vacuum energy density is greater than the sum of the radiation density plus one-half the matter density. However, acceleration via the cosmological constant has a seriously unattractive feature. The vacuum energy density is a constant and so would have had to be delicately fine-tuned in the early Universe to give an energy density today that has about the same order of magnitude as the matter and radiation densities. Furthermore, that energy density is far lower than expected from theoretical estimates (see chapter 8). This is a long-standing puzzle known as the *cosmological constant problem*.[5]

By contrast, a variable energy density could have started high and fallen off as the universe expanded, just as presumably happened for the matter and radiation densities. In fact, one can expect these three densities to more or less track each other as the universe expands, if they interact even weakly. Currently the cosmic microwave background has largely decoupled from matter and the radiation density is negligible compared to the other density components. However, the dark energy may be more closely coupled to matter. Let us look at the model in which the dark energy is contained in a scalar field.

COSMIC ACCELERATION VIA A SCALAR FIELD

Another way to achieve gravitational repulsion is for the dark energy to comprise a physical medium with negative pressure. In this case, the pressure need not precisely equal minus the energy density. When the density is not constant, it can evolve along with the other components—radiation and matter—as the Universe expands. This still-unidentified substance has been dubbed *quintessence* and would constitute the dark energy that makes currently up 70 percent of the mass-energy of the Universe.[6]

Let us consider the case where quintessence constitutes the dark energy field. In quantum field theory, that field will be composed of particles—the quanta of the field. In the simplest case, the field associated with the vacuum is what is called a *scalar field*. (By contrast, the electric and magnetic fields are *vector fields*). The quantum of a scalar field has zero spin. In this picture, these quanta are viewed as another component to the Universe besides matter and radiation—one with negative pressure that results in a gravitational repulsion between particles. Since the density can vary with time, this could solve the cosmological constant problem mentioned earlier. Negative pressure for a boson field is not ruled out, and in fact may be quite natural since bosons have a tendency to attract one another in quantum mechanics—just as fermions tend to repel.

In classical field theory, it can be shown that the equation of motion of a scalar field is mathematically equivalent to the equation of motion of a unit-mass nonrelativistic particle moving in one dimension. In an expanding universe, the equation of motion for the field contains a "friction" or damping term that results from the expansion of the universe.

If the Universe is assumed to be dominated by a scalar potential, then its average history is fully computable, depending only on the assumed initial conditions and the parameters of the model.

AT THE PLANCK SCALE

If we extrapolate the expanding Universe backward in conventional time, we find we cannot go all the way to zero time. The smallest operationally definable time interval is the *Planck time*, which is on the order of 10^{-43} second. Likewise, the smallest operationally definable distance is that traveled by light in the Planck time, 10^{-35} meter, which is called the *Planck length*. We can also define a *Planck mass*, which is on the order of 10^{-10} gram, and a *Planck energy*, which is on the order of 10^9 Joules or 10^{28} electron-volts. The Planck mass and energy represent the uncertainties in rest mass and rest energy within the space of a Planck sphere or within a time interval equal to the Planck time. Note, however, that unlike the Planck length and time, the Planck mass and energy are not the smallest values possible for these quantities. In fact, they are quite large by subatomic standards.

Although time zero is specified as the time at which our Universe began, the smallest operationally defined time interval is the Planck time, 10^{-43} second. Within a time interval equal to the Planck time, the Heisenberg uncertainty principle says that the energy is uncertain by an amount on the order of the Planck energy. That energy will be spread throughout a sphere with a radius on the order of the Planck length. The energy density within that sphere, for this brief moment, will be on the order of $(10^{28})/(10^{-33})^3 = 10^{121}$ electron-volts/cm^3, which we can call the *Planck density*. This density would be equivalent to a cosmological constant.

Note that the minimum space and time intervals defined by the Planck scale apply to every momentum in time and position in space, not just the origin of the Universe. No measurements can be made with greater precision, no matter where or when.

THE ENTROPY OF THE UNIVERSE

Clearly a certain amount of order exists in the cosmos, in the structure of galaxies, stars, and planets, even though, as I have pointed out, that order is contained in only a half of 1 percent of the mass of the Universe. Still, for that structure to be assembled from a less orderly system, such as a chaotic ball of gas, entropy had to be removed from the system. This in itself does not violate the second law of thermodynamics, as long as the entropy removed is compensated for by the increase in the entropy of the surroundings by at least the same amount.

The second law does not prevent a nonisolated system from

becoming more orderly. But what about the Universe as a whole? Presumably it constitutes an isolated system, at least as a working hypothesis. Did the Universe not have to begin in a more orderly state than it is now, since disorder, or entropy, increases with time? Does this not require that the Universe was formed with preexisting order?

This indeed was a problem when it was thought that the Universe had a fixed volume, that is, was a "firmament" in which the stars were more or less fixed with respect to one another. However, as Hubble discovered, the Universe is not a firmament. It is an expanding gas that is not in the state of thermal equilibrium for which its entropy would be maximal. As it increases in volume, more particle states become accessible and the Universe's maximum allowable entropy increases further. In fact, the Universe could have begun in a state of maximum entropy—complete chaos—and still had local pockets of order form as it expanded from an original state of complete disorder.

How far is the current Universe away from equilibrium? First, let us estimate its maximum possible entropy. Second, let us estimate its current total entropy.

Since we cannot see inside a black hole, we have no information about what is inside and can conclude that it has the maximum entropy of any object its size. It follows that the maximum entropy of the Universe is that of a giant black hole of the same size. That maximum entropy is estimated to be 10^{122} for the visible Universe.[7]

The visible Universe is not one giant black hole. However, it most likely does contain many "smaller" black holes, especially in the cores of galaxies. Roger Penrose has argued that the total entropy of the Universe should be dominated by the sum of the entropies of all the black holes within, which he estimates at 10^{100}. Based on this estimate, currently we have at least 22 orders of magnitude of room available for order to form.

When we describe the history of the particular Universe within which we reside in terms of observations we make from the inside, we must begin at the Planck time with a sphere of radius equal to the Planck length. Because of the uncertainty principle, the energy within this volume cannot be known within an amount on the order of the Planck energy. This implies that the mass enclosed is uncertain by the order of the Planck mass. Now, a body of Planck mass inside a sphere of radius equal to the Planck length is a black hole.[8] That is, the Universe at the

Planck time is indistinguishable from a black hole, which, in turn, implies that the entropy of the Universe was maximal at that time.

We can thus conclude that our Universe began in total chaos, with no order or organization. If any order existed before that time, it was swallowed up by the primordial black hole.

Why did the Universe not stay a Planck-sized black hole? In the 1970s Stephen Hawking showed that black holes are unstable—that they eventually disintegrate. He calculated that the average lifetime of a black hole is proportional to the cube of its mass. For the black holes that form from collapsing giant stars, this lifetime will be far greater than the age of the Universe. However, a Planck-sized black hole will disintegrate on the order of the Planck time, 10^{-43} second.

We thus have the picture of the Universe starting out at the Planck time as a kind of black hole of maximum entropy—with *no structure or organization*. It then explodes into an expanding gas of relativistic particles. Let us forgo for the moment any discussion of where that black hole came from and consider what happened after it exploded.

ROOM FOR ORDER AFTER THE PLANCK TIME

The Planck energy is the uncertainty in energy for a sphere of Planck dimensions. No smaller energy can be measured under those conditions and so we must assume the sphere can contain energy of about this order of magnitude. It seems likely that this energy will remain in the form of kinetic energy for at least a short while after the Planck time. That is, the particles that emerge from the explosion will be in the form of radiation, relativistic particles of negligible rest energy. The maximum entropy of the radiation will be less than that of a black hole of the same size. As the radiation expands, its entropy increases, but not as rapidly as the maximum allowable entropy. This allows increasing room for order to form in the early Universe. In other words, after the Planck time the expanding Universe opens up increasing numbers of accessible states, allowing the particles to become more organized by natural processes. While we do not know the details of those processes, it should at least be evident that the Universe could have begun in maximum disorder and yet produced order without the violation of the second law or any other principle of physics.

This provides an explanation for the question raised in the last chapter—why is the current Universe not in equilibrium? Of course, if it were we would not be here to talk about it. Our Universe was once in equilibrium—when it was a Planck-sized black hole. This equilibrium was unstable, however, able to hold together for only 10^{-43} second.

INFLATION

As already mentioned, in the early 1980s, well before the discovery that today's Universe is accelerating, it was proposed that during the first 10^{-35} second or so, the early Universe underwent a very rapid, accelerating exponential expansion, called inflation, before commencing with the Hubble expansion. Inflation solved a number of outstanding problems in the previous big-bang cosmology that I will only briefly mention since they are extensively covered in both the popular and the technical literature.[9]

1. The Horizon problem

The cosmic microwave background radiation has a spectrum that is characterized by a black body of temperature 2.725 degrees Kelvin, with fluctuations of one part in 10^5. This is true for regions of space that could not have been causally connected in the early Universe if one simply extrapolates back the Hubble expansion. In the inflationary model, these regions were much smaller and in causal contact when the radiation was produced.

2. The Flatness problem

Observations indicate that today's Universe is very close to being spatially geometrically flat (Euclidean) on average (that is, a Euclidean-space model fits the data). The problem is to explain why. Inflation does that trivially since it asserts that the Universe has expanded by many orders of magnitude since the early Universe, and so it is expected to be very flat, like the surface of a balloon that has been blown up to huge proportions. Note, however, that this does not necessarily imply that the Universe is *exactly* flat or that its average density *exactly* equals the critical density.

These could be very close approximations. In particular, a closed Universe is not ruled out, in which many billions of years in the future, the expansion eventually stops, gravitational attraction takes over, and Universe begins to collapse.[10]

3. The Monopole problem

Elementary particle physics implied that the Universe should be filled with *magnetic monopoles*, point magnetic charges that resemble point electric charges, where the field radiating from the point is magnetic rather than electric. Inflation reduced the number to about one in the visible Universe, consistent with the fact that no monopoles have ever been observed.

4. The Inhomogeneity problem

In the noninflationary big bang, thermal fluctuations in the early Universe should have resulted in a much lumpier universe than we have. In the inflationary Universe, the fluctuations occur when the Universe was much smaller leading to a smoother Universe. Indeed, the inflationary model required that the cosmic microwave background should show a specific pattern of temperature fluctuations across the sky. This pattern has been confirmed by observations. The failure to see this pattern would have doomed the model.

Indeed, observations of ever-increasing precision have remained beautifully consistent with theoretical calculations using the inflationary model, which, until recently, has had no serious competition. As we have discussed, the expansion of Universe is currently accelerating. This has suggested that perhaps this current inflationary epoch is sufficient to solve the above problems without early Universe inflation. In one scenario, called the *cyclic Universe*, the Universe repeats cycles of expansion and contraction.[11] The notion of an oscillating universe was first proposed many years ago but discarded because of the entropy production in each cycle. The new model claims to avoid this.

The true (that is, best) scenario may become clear in the near future, as new and better experiments on Earth and in space establish the nature of the dark matter and, in particular, the dark energy. These could lead to surprises that require us to rethink our models, from subnuclear physics to cosmology. So far, however, no drastic revision in these models is implied by the data. As remarkable and unexpected was the discovery that the Universe is currently accelerating, the theoretical basis for this has already existed for almost a century in general relativity. If anything, this discovery has helped to solidify existing models such as inflation

NOTES

1. For a popular-level review, see Timothy Ferris, *The Whole Shebang: A State-of-the-Universe Report* (New York: Simon & Schuster, 1997).

2. Alan Guth, "The Inflationary Universe: A Possible Solution to the Horizon and Flatness Problems," *Physical Review* D23 (1981): 347–56; *The Inflationary Universe* (New York: Addison-Wesley, 1997).

3. A. Reiss et al., "Observational Evidence from Supernovae for an Accelerating Universe and a Cosmological Constant," *Astronomical Journal* 116 (1998): 1009–38; Perlmutter, S., et al. "Measurements of Omega and Lambda from 42 High-Redshift Supernovae." *Astrophysical Journal* 517 (1999): 565–86.

4. Other plausible explanations than dark energy for the cosmic acceleration have been recently proposed. See Edward W. Kolb et al., "Primordial Inflation Explains Why the Universe Is Accelerating Today," http://www.arxiv.org/abs/hep-th/0503117 (accessed March 14, 2005); David L. Wiltshire, "Viable Exact Model Universe without Dark Energy from Primordial Inflation," http://www.arxiv.org/abs/gr-qc/0503099 (accessed March 23, 2005).

5. S. E. Rugh and Henrick Zinkernagel, "The Quantum Vacuum and the Cosmological Constant Problem," *Studies in History and Philosophy of Modern Physics* 33 (2001): 663–705.

6. Jeremiah P. Ostriker and Paul J. Steinhardt, "The Quintessential Universe," *Scientific American* (January 2001): 46–53.

7. Entropy is dimensionless in units where Boltzmann's constant $k = 1$.

8. Some might dispute this claim, but it depends on the definition of a black hole. I am using the standard definition in which the radius of a body is less than its Schwarzschild radius, that is, the radius at which the escape velocity exceeds the speed of light.

9. D. Kazanas, "Dynamics of the Universe and Spontaneous Symmetry Breaking," *Astrophysical Journal* 241 (1980): L59–63; Guth, "The Inflationary Universe: A Possible Solution to the Horizon and Flatness Problems"; Guth, *The*

Inflationary Universe; André Linde, "A New Inflationary Universe Scenario: A Possible Solution of the Horizon, Flatness, Homogeneity, Isotropy, and Primordial Monopole Problems," *Physics Letters* 108B (1982): 389–92.

10. Marc Kamionkowski and Nicolaos Toumbas, "A Low-Density Closed Universe," *Physical Review Letters* 77 (1996): 587–90.

11. Paul J. Steinhardt and Neil Turok, "A Cyclic Model of the Universe," *Science* 296 (2002): 1436–39.

CHAPTER EIGHT

OUT OF THE VOID

THE PHYSICS OF EMPTINESS

In the last chapter we discussed the inflationary scenario that describes an early universe expanding exponentially during the first tiny fraction of a second after the Planck time. In this cosmological model, our Universe begins in a highly symmetric and maximum entropy state at the Planck time—the earliest time that can be operationally defined. If this was indeed the case, then the Universe was without structure at that time. We saw that the appearance of structure after the Planck time did not violate the second law of thermodynamics, since as the Universe expanded, its entropy did not increase as fast as its maximum allowable entropy, thus allowing increasing room for order to form. However, we still can ask: Where did this structure come from?

Suppose we have a container empty of matter. We call the state inside the container a *vacuum*. In familiar parlance, a vacuum has zero energy. However, in physics we define the vacuum as the lowest energy state of a system empty of matter, allowing for the possibility that it may still contain something we can identify as "energy." In particular, general relativity allows for the possibility that energy may still be present in an otherwise empty universe, stored in the curvature of space. The energy density in this case is proportional to the cosmological constant. And, importantly, it is the same for all times—from the beginning of the Universe to its end. A universe with only a cosmological constant—energy density is still a vacuum, as I have defined the term.

We saw that a cosmological constant may be responsible for the current acceleration of the Universe. That is, this is one possible source of the gravitationally repulsive dark energy, which apparently constitutes 70 percent of the average energy density of the Universe. Alternatively, the dark energy may be stored in a yet unidentified field that has negative pressure. This hypothetical field has been dubbed *quintessence*. Such a field, described quantum mechanically, would be composed of particles that constitute the quanta of the field. A universe that is pure quintessence would thus not be a vacuum.

Let us attempt to develop a physics of the vacuum, that is, a physics of a system in which all the matter, including dark matter and quintessence, has been removed. It may strike the reader as an empty task, but, as we will see, we can still describe the properties of an empty universe using well-established physics.

Neither an empty universe nor a universe containing some form of uniformly distributed energy or matter has anything in it to define an absolute position or direction in space-time. Thus, the space-time symmetries we have already discussed would be in effect. You might argue that a truly empty universe has no space and time either. That does not negate but further supports the statement that no special position, time, or direction exists. Besides, space and time are human inventions. Once you introduce the space-time model to describe observations, the laws of conservation of energy, momentum, angular momentum, and special and general relativity apply for any physical description you might make for an empty universe, and any matter that then appears would be expected to follow these principles in the absence of any external agent to specify otherwise.

We can reasonably hypothesize that gauge invariance will also apply to any description of the empty universe in its simplest form. That is, any law of physics we choose to write down cannot depend on the orientation of the state vector in the abstract space we use to describe that state. Quantum mechanics and all of the global principles of physics would be included in our theoretical models. And once we have quantum mechanics, we have the plausibility of a spontaneous, indeterministic mechanism for the transition between different states that cannot happen in classical physics. This includes a transition from a state of emptiness to one of matter, normal or abnormal.

So, what are the properties that physics predicts for an empty universe? To answer this we need to start with a nonempty universe and ask what happens when the matter and radiation inside are removed.

ZERO-POINT ENERGY

In classical field theory, the equation of motion of a field is the same as that of a particle of unit mass attached to a spring. This is a mathematical equivalence, and should not be taken to imply that some mass is actually oscillating. It simply means that we can use the mathematical methods of particle mechanics as applied to the harmonic oscillator to describe the behavior of fields. This comes in particularly handy when we want to consider quantum fields. They will behave mathematically like quantum harmonic oscillators, which have been well understood theoretically since the early days of quantum mechanics. This enables physicists to apply these established mathematical results to quantum fields.

The quantum oscillator has a very simple structure of energy levels separated by a fixed energy. These levels can be thought of as forming a ladder with equally spaced rungs, as shown in fig. 8.1. The higher you go up the ladder, the higher the energy of the oscillator.

When applied to quantum fields, each energy level corresponds to a given excited state of the field, which can be described as containing a certain number of quanta. That is, we view the field as if it were a field of particles—just as a beach is a field of sand pebbles.

In particular, the electromagnetic field is not some tension in an ethereal medium, as was thought in the nineteenth century. Instead, in the quantum view, the field is a region of space filled with photons. A monochromatic beam of light, such as produced by a laser, is composed of equal-energy photons. As more energy is pumped into the beam, the number of photons increases proportionately. The energy level of the field moves up one rung on the ladder for every new photon produced.

Similarly, if we have a field containing a certain number of quanta, we can move down the ladder one rung at a time by removing one quantum at a time. After all the quanta have been removed and we arrive at the bottom rung, we find that the lowest energy level or "ground state" energy is not zero. In fact, it has an energy equal to one-half the energy difference between levels. In other words, a quantum field with zero quanta still has some energy, which is called the *zero-point energy.*

The zero-point energy of a common oscillating body, such as a mass on a spring, can be understood as an effect of the uncertainty principle. To have exactly zero energy, that body has to be exactly at rest, which means its momentum has to be zero with vanishing uncertainty. However, according to the uncertainty principle, in order to have zero uncertainty in momentum, the particle must have infinite uncertainty in position. As you better localize a body in space, you increase the probability that when you measure its momentum you will not get zero. Indeed, this is true for any

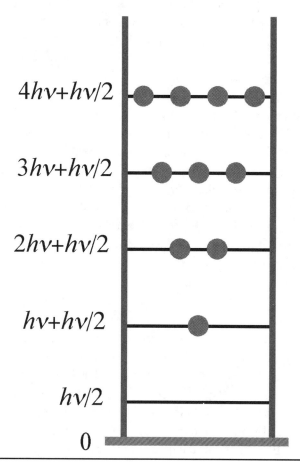

Fig. 8.1. The energy levels of a quantum harmonic oscillator form a ladder with equally spaced rungs. The spacing between levels is $h\nu$, where h is Planck's constant and ν (Greek *nu*) is the oscillator frequency. The lowest energy state has energy $h\nu/2$. This is the zero-point energy of the oscillator. When applied to a radiation field, each quantum has energy $h\nu$. The spheres on the various levels indicate the number of quanta at the level. The ground state contains no quanta.

particle, not just oscillators. As long as a particle is localized, even if that localization is no better than simply somewhere in the visible Universe, you will generally not measure that particle to be at rest. The energy of that minimal motion is the zero-point energy.

However, this explanation does not suffice for the zero-point energy of quantum fields. If the lowest energy level is a state of zero quanta, what carries the energy? One common explanation is that the zero-point energy is carried by particle-antiparticle pairs flitting in and out of the vacuum as the result of "quantum fluctuations."

Short-lived electron-positron pairs are known to be important in producing effects such as vacuum polarization in quantum electrodynamics. However, these pairs are produced by photons and annihilate back into photons of nonzero energy. Pairs that appear spontaneous in the vacuum have zero energy and so, it would seem, cannot be responsible for nonzero zero-point energy.

A theoretical estimate of the density of zero-point energy in the Universe, left over after all the photons have been removed, is 10^{124} electron-volts/cm^3. This is to be compared with the observational fact that the energy density of the Universe is very close to the critical density, which means that the zero-point-energy density can be no more than on the order of 10^4 electron-volts/cm^3. Surely the above theoretical estimate must be one of the worst calculations in the history of physics, disagreeing with the data by 120 orders of magnitude! Indeed, if the zero-point energy were as dense as calculated, the curvature of the Universe would be so great that you would not be able to see your hand in front of your face—assuming your hand and face could stay together in the presence of such an enormous gravitational repulsion.

As mentioned, this discrepancy is usually referred to as the cosmological constant problem, based on an association of the zero-point energy with the cosmological constant.[1] We could also call it the *vacuum energy problem*. Even if the cosmological constant were intrinsically zero, a nonzero vacuum energy is implied by standard physics. Surely the above calculation is wrong. Other estimates have been made that are not as far off, but still in substantial disagreement with the data, by orders of magnitude.

When gravity is not considered, the zero-point energy actually has no effect since, in that case, energy is defined within an overall arbitrary constant that can be subtracted away. However, in general relativity the absolute magnitudes of the terms in the energy-momentum tensor determine the curvature of space-time. Even in the absence of matter and radiation, any zero-point energy will have a gravitational effect.

POSSIBLE SOLUTIONS TO THE VACUUM ENERGY PROBLEM

So, what is the energy density of the vacuum? We have noted that the zero-point energy of a harmonic oscillator is half the energy level difference. This quantity is positive for a boson (integer spin particle) such as the photon, but *negative* for a fermion (half-integer spin particle) such as the electron. Thus, a small or even zero vacuum energy is not inconceivable when we properly include both boson and fermion fields, since their zero-point energies tend to cancel. If the vacuum is supersymmetric, the above cancellation is exact since supersymmetry implies an equal number of bosonic and fermionic degrees of freedom. That is, the supersymmetric vacuum has exactly zero energy.

However, the fact that we have not observed the supersymmetry partners of familiar particles at exactly the same masses indicates that supersymmetry is broken in the cold Universe in which we live. If the current vacuum is not supersymmetric, some net zero-point energy may still be present. Theoretical estimates that attempt to take this into account are still orders of magnitude higher than the observed upper limit, and so do not solve the vacuum energy problem. Nevertheless, it may be that the vacuum energy remains exactly zero, for reasons still unknown, and what is being interpreted as vacuum energy, the dark energy responsible for the acceleration of the Universe, is in fact a material energy such as quintessence.

Robert D. Klauber and J. F. Moffat have independently suggested another solution to the vacuum energy problem.[2] The equations for a pure radiation field allow for negative-energy solutions as well as those of positive energy. If we include those solutions, the zero-point energy exactly cancels.

The quanta of this field are negative-energy photons. How might we interpret them? In the late 1920s Dirac interpreted the negative energy solutions of his relativistic equation describing electrons as antielectrons or positrons. Here we might do the same, namely, to include the presence of "antiphotons," which, in this case, are empirically indistinguishable from photons.

The same results obtain if instead of dealing with negative energies we change the direction of time, that is, have positive-energy photons moving backward in time. In doing so, we simply recognize, as I have several times in this book and my earlier work *Timeless Reality*,[3] the inherent time symmetry of physics. As with a number of the so-called quantum paradoxes, we find that a problem goes away when you simply do not force your personal prejudice of directed time on the physics, but let the equations speak for themselves.

THE CASIMIR EFFECT

Nevertheless, the existence of a small vacuum energy is indicated by certain empirical observations, in particular, the *Casimir effect.* In 1948 Hendrick Casimir argued that if you had two conducting plates separated in a vacuum, a net force pulling the two plates together would exist.[4] Steven Lamoreaux confirmed the effect in the laboratory in 1997.[5] This force is explained by noting that fewer states are available to photons in the gap between the plates than those outside. Because the plates are conducting, the electric field is zero on their surface, providing a boundary condition so that the maximum wavelength of the field is equal to twice the plate separation. Since no such limitation exists outside the gap, an energy imbalance results in the plates being pushed together.

So, how can we reconcile the apparent presence of vacuum energy with the possibility that the zero-point energy of a totally empty vacuum, which I call a *void*, is exactly zero? M. Borgag, U. Mohideen, and V. M. Mostepanenko have given a possible answer in their extensive review of the Casimir effect.[6] They point out that the Casimir effect involves material boundaries, plates that can polarize the vacuum. The phenomenon of vacuum polarization is well understood and results in nonzero vacuum energy. Thus, the Casimir effect is not the result of any zero-point energy in an empty vacuum, but a vacuum polarization that results from the nearby presence of matter. R. L. Jaffe concurs with this conclusion in a recent paper.[7]

THE LAWS OF THE VACUUM

A vacuum is obtained by taking a container of particles and removing all the particles. The mathematical procedure is implied by the quantum mechanics of free bosonic and fermionic fields.

Although the vacuum state has no particles, it remains a legitimate object for our abstract theoretical description. A state with n particles is well defined. Any principles we apply to a state with n particles can be legitimately applied to a state with $n = 0$. The methods of quantum field theory provide the means to move mathematically from a state with n particles to a state of more or fewer particles, including zero particles.

As we have seen, when a state is invariant under a particular coordinate transformation, then the generator of that transformation is associated with an observable that is conserved (Noether's theorem). We can use the symmetries of the vacuum state to determine the "laws" that apply to the vacuum. Not all will necessarily apply when the system is no longer a vacuum, but we can hypothesize that the laws that do apply to the vacuum might still apply when particles appear spontaneously out of the vacuum. And those laws that may not apply may have arisen by an equally spontaneous, noncausal breaking of the symmetry of the vacuum.

Since the vacuum, as described above, has an unknown energy density, let me use the term *void* to describe a vacuum that has zero vacuum energy and is not confined to a small region of space as in the Casimir effect. We have seen that zero vacuum energy occurs for a supersymmetric vacuum or where we include the negative-energy solutions for the radiation field. That is, its total energy is strictly zero, including any cosmological constant. If the Universe at its earliest moments were totally chaotic and symmetric, then it would be a void with zero energy. Thus there should be no cosmological constant problem in the early Universe.

The next issue is how we can get a universe of matter from the symmetric void.

TUNNELING THROUGH

Surprisingly, we can also examine this question theoretically. In classical physics, when a barrier exists between two states of a system and bodies have insufficient energy to mount the barrier, then the states on either side of the barrier remain forever out of contact. For example, if you have a dog in an enclosed yard, as long as the surrounding fence is high enough, the dog will remain confined to the yard. In quantum mechanics, however, a nonzero probability exists for an object to penetrate a barrier by what is called *quantum tunneling*. The penetration probability is simply very small for macroscopic objects like dogs.

Consider a nonrelativistic particle incident on a square potential barrier as shown in fig. 8.2. The wave function for the particle on the left of the barrier describes a free particle of a certain momentum, which can be measured as a real number—the mass multiplied by the velocity. Quantum mechanics allows for a nonzero wave function inside the barrier. However, the particle described by this wave function has an *imaginary* momentum, that is, the mathematical quantity that represents momentum in the equations contains a factor $\sqrt{-1}$. Its square is negative. Since all observed particles have momenta that are given by real numbers, by operational definition, we say that the wave function inside the barrier describes a situation that is "nonphysical." Nevertheless, these nonphysical wave functions are mathematically allowed solutions of the Schrödinger equation.

The wave function to the right of the barrier will again be that of a free particle with real momentum. The probability for transmission through the barrier is derived in many textbooks and has been verified empirically. In fact, quantum tunneling is a well-established phenomenon, accounting quantitatively for nuclear alpha decay. It also forms the basis for the scanning tunneling microscope (STM), a powerful instrument that enables physicists to take pictures of individual atoms. The model of quantum tunneling works—spectacularly.

Quantum tunneling provides a beautiful example of how the equations of physics can be extended meaningfully to nonphysical (unmeasurable) domains. One often hears that science should concern itself only with what is directly observable. Here we have a situation in which the wave function inside the barrier describes a particle with nonphysical, imaginary momentum. Such a particle is unobservable. Yet taking its mathematical representation seriously enables us to make calculations of tunneling probabilities that can be compared with real measurements, such as the decay rates of nuclei.

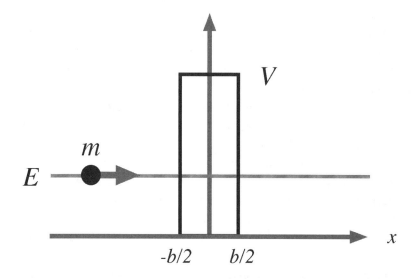

Fig. 8.2. A particle of mass m and energy E incident on a potential energy barrier of height V and width b. Quantum mechanics allows for particles to tunnel through the barrier with a calculable probability. The region inside the barrier is "unphysical," that is, unmeasurable, with the particle's momentum being an imaginary number.

Note that such an approach is not inconsistent with the operational view that has been espoused in this book, where quantities are defined by how they are measured and the equations of physics are developed to describe those measurements. Here we have operational quantities in a physical domain, such as momentum and energy, being described by equations, such as the Schrödinger equation, developed from observations. We then mathematically interpolate those equations through a nonphysical domain so that two physical domains can be connected. This enables us to make calculations that are still testable in the physical domains.

A number of authors have suggested that the Universe appeared as a quantum tunneling process from a previous nonphysical domain.[8] What is that domain? A reasonable guess is the Planck-sized void prior to the Planck time, where no quantities can be operationally defined.

As I have argued, time is fundamentally symmetric.[9] Thus, while we cannot make measurements within a Planck time interval centered around the point in time we define as $t = 0$ (or any time, for that matter, even "now"), all the equations of physics that apply for $t > 0$ apply equally well for $t < 0$. In particular, the same Einstein equation for an empty universe with a cosmological constant that leads to an exponential inflation for $t > 0$ will give an exponential deflation for $t < 0$.

However, time's arrow is defined by the direction of increasing entropy. Thus, we need to reverse the direction of time on the other side of the time axis so that we have inflation in both directions away from $t = 0$.

Note that this serves to answer one of the puzzles of cosmology: Why, if time is symmetric, do we have such a pronounced arrow of time in our own experience?[10] There is an equally pronounced arrow of time in the other direction, on "the other side of time," so it is a "local" effect with time symmetry still valid globally.

So, from our point of view, we can imagine an incoming wave function on the negative side of the time axis hitting a potential barrier and tunneling through the intermediate void to give our Universe.[11] This is basically the "no boundary" model proposed by James Hartle and Stephen Hawking back in 1983, although they did not consider time symmetry.[12] Hawking, however, has remarked that the no-boundary quantum is CPT invariant and so "there will be histories of the Universe that are the CPT reverses" of what is normally described in the model.[13] (See chapter 5 for a discussion of CPT).

I know I will be criticized for speculating about things that we can never observe, like a mirror universe in our past. But, I am doing nothing more than pointing out the implications of the current principles of physics and cosmology that describe our observed Universe so well. Those principles, we have seen, largely arise out of simple symmetry considerations that would apply equally to either side of the time axis. The universe on the other side of our time axis would be expected to have the same global laws, such as the great conservation principles, and be described by the same quantum mechanics and relativity that work so well in our Universe. Those universes might be expected to have gravity and electromagnetism as we know it, with massless photons a major component.

However, our mirror universe would not be expected to duplicate every detail of the Universe on our side of the time axis. Those structural elements like quarks, leptons, and gauge bosons that result from spontaneous broken symmetry would be expected to be different.

We have a situation that becomes very difficult to describe in conventional, tense-based language. The "origin" of our Universe is at $t = 0$, but this is also the origin of a universe in our past, which evolves, in thermodynamic time opposite to our thermodynamic time. And, if this happened at the arbitrary point in space-time that we have chosen as the origin of our coordinate system, why could it not have happened at other places as well?

NOTES

1. Steven Weinberg, "The Cosmological Constant Problem," *Reviews of Modern Physics* 61 (1989).

2. Robert D. Klauber, "Mechanism for Vanishing Zero Point Energy," http://www.arxiv.org/pdf/astro-ph/0309679 (accessed January 20, 2005); J. W. Moffat, "Charge Conjugation Invariance of the Vacuum and the Cosmological Constant Problem," http://xxx.lanl.gov/abs/hep-th/0507020m (accessed July 21, 2005).

3. Victor J. Stenger, *Timeless Reality: Symmetry, Simplicity, and Multiple Universes* (Amherst, NY: Prometheus Books, 2000).

4. H. B. G. Casimir "On the Attraction between Two Perfectly Conducting Plates," *Proceedings Koninkl. Ned. Akad. Wetenschap.* 51 (1948): 793–95.

5. S. K. Lamoreaux, "Demonstration of the Casimir Force in the 0.6 to 6 μM Range," *Physical Review Letters* 78 (1997): 5–8; 81 (1997): 5475–76.

6. M. Bordag, U. Mohideen, and V. M. Mostepanenko, "New Developments in the Casimir Effect," *Physical Reports* 353 (2001):1–205.

7. R. L. Jaffe, "The Casimir Effect and the Quantum Vacuum," http://www.arxiv.org/pdf/hep-th/0503158 (accessed March 21, 2005).

8. A. Vilenkin, "Boundary Conditions and Quantum Cosmology," *Physical Review* D33 (1986): 3560–69; André Linde, "Quantum Creation of the Inflationary Universe," *Lettre al Nuovo Cimento* 39 (1984): 401–405; David Atkatz and Heinz Pagels, "Origins of the Universe as Quantum Tunneling Event," *Physical Review* D25 (1982): 2065–73.

9. Stenger, *Timeless Reality.*

10. Huw Price, *Time's Arrow and Archimedes' Point: New Directions for the Physics of Time* (Oxford: Oxford University Press, 1996).

11. For a pre–big bang scenario suggested by string theory, see M. Gasperini and G. Veneziano, "The Pre–Big Bang Scenario in String Cosmology," *Physics Reports* 373 (2003): 1–212.

12. J. B. Hartle and S. W. Hawking, "Wave Function of the Universe," *Physical Review* D28 (1983): 2960–75.

13. S. W. Hawking, "Quantum Cosmology," in *Three Hundred Years of Gravitation*, ed. S. W. Hawking and W. Israel (Cambridge: Cambridge University Press, 1987), p. 649.

CHAPTER NINE

THE COMPREHENSIBLE COSMOS

WITHIN OUR GRASP

When we look at the sky at night, we see thousands of stars, each one a sun not too different from our own. Peering through a telescope reveals many more stars and galaxies of stars. From the data gathered with their instruments, astronomers have estimated that our Sun is one of approximately 100 billion stars in a galaxy called the Milky Way. Beyond our galaxy lie 100 billion more galaxies, each with 100 billion or so stars, reaching out some 10 billion light-years or more. And this is just what lies within our visible horizon, the distance beyond which we cannot reach, in the age of the Universe, with signals traveling at the speed of light. Modern cosmological models suggest a much vaster universe lies beyond that horizon, forever beyond our observation, and these models allow the possibility of many other universes as well.

Back on Earth, oceans, clouds, mountains, deserts, and a dizzying collection of plants and animals form the panorama that greets our eyes. And, once again, our scientific instruments reveal far more—a submicroscopic world of tiny particles inside matter, which bear at most a superficial resemblance to the material bodies of common experience.

The immense size and complexity of the Universe and the strangeness of the microworld, have led laypeople and scientists alike to assume that nature must forever remain largely shrouded in mystery. Surely, it is presumed, the finite intellect of human beings cannot be expected to grasp more than but a tiny portion of the data that flood our senses and instruments.

In this book I have challenged that assumption. The natural world—the cosmos that is currently revealed by our observations—is comprehensible. Far from representing unfathomable mysteries, existing physics and cosmology describe a universe that is well within the grasp of our intellect. Let me summarize what we have learned and attempt to draw some general conclusions.

ILLUMINATING OBJECTS

The great success of physical models strongly implies that they are more than simply useful descriptions of observations. These models are too full of surprises and amazing predictions of otherwise unanticipated future events to be arbitrary and purely culturally determined. Surely physical science is saying something about ultimate reality. But, as I have emphasized, the models that we use in describing observations need not exactly mimic whatever reality underlies those observations. If they do not, they still can be valuable. As long as a model describes all observations pertinent to its domain, then it can be used to predict future observations within that domain.

In the next chapter I will speculate on the nature of that reality, but it may disappoint the reader to find that we have at least two alternative "metaphysical models" that cannot be distinguished by the data. The model that I prefer, because it is the simplest, is a reality of composed of elementary localized objects, or "atoms," moving about in an otherwise empty void. I call this *atoms and the void*. To me this is a comprehensible cosmos.

However, I have to admit that another viable metaphysical model exists that is probably favored among mathematicians and some theoretical physicists. This is a Platonic model in which reality is composed, not of grubby electrons and quarks, but of quantum fields that live not in space-time, but in the abstract space of mathematical "forms." To most people, including myself, this would be an incomprehensible cosmos.

But before getting into those speculations, let us look at what we can comprehend with some confidence. In science we gain knowledge, first of all, by way of observation. I have attempted to provide a simple characterization of the act of observation as a process of kicking and measuring what is kicked back. By "kicking" and "kicking back" I refer more precisely to the process of energy and momentum transfer that occurs in any act of observation, whether by eye or by instrument. Matter is then defined, in the models used in physics, as stuff that emits or reflects energy and momentum from source to observer. This matter is further visualized as distinguishable objects that can be uniquely identified and labeled by the act of measurement and the data obtained in that act. The model further hypothesizes the properties of these objects and attempts to infer predictions that can be tested against experiment and observation.

Now, this would seem to eliminate any immaterial ingredient to reality—by definition. Indeed, one often reads that science is limited to explaining phenomena materialistically. However, both views are mistaken. Of course, if a purely material picture is sufficient to provide an adequate description of all observations, then no additional ingredients are required as part of the model and should not be included. If, on the other hand, a purely material description were to prove inadequate, then a rationale would exist for introducing the hypothesis of some additional element into our model. In that case, nature would include an immaterial component. So far, including the immaterial has not proved necessary.

In the most common form of observation, again as described in the space-time model, light from the Sun or other sources strikes an object and is reflected into our eyes. Or the object may emit light. While the latter process may not be a direct reaction to the act of an observer, it still involves momentum and energy transfer. Objects can kick us without us kicking them in the first place. Our other senses—sound, smell, and touch—involve analogous processes.

In controlled experiments, we illuminate the objects of our study with photons or other probing particles that carry calibrated momentum and energy, enabling us to quantify our observations. Thus, the elementary measurement process can be viewed as the simple exchange of momentum and energy between the observed and the observer. With scientific instruments, we extend our ability to probe reality so that photons outside the visible spectrum and other particles, such as neutrinos, can be used.

Scientific observations are generally more precise and quantitative than the everyday acts of human observation, but the latter can be quantitative as well. For example, in normal life we often note the time on a clock at which an event occurs and locate it in space.

In science an atomic clock is used to define a time standard against which other clocks are synchronized. According to current convention, space—specifically the distance between two points—is defined as the time it takes for light (in a vacuum) to move between those points. In chapter 2 we discussed using a clock to time pulses of light that we send out into an otherwise unobserved universe. The timing of any returned signals is then used to specify the position, velocity, acceleration, and other physical variables of the objects that we hypothesize are "existing out there" and sending signals back to us. Note that this space-time description is a model invented to describe those observations. We have no other way of "seeing" the objects so described and no way of verifying that reality is "truly" composed of those objects or that the terms distance, velocity, and acceleration are "true" elements of reality. We can only argue that this model is simple and consistent with the data. As long as it agrees with all the data, we are not being irrational to apply it as far as we can.

The space-time model is not our sole option for describing the data. In the 1960s physicists attempted to eliminate space and time from their theoretical schemes and developed a model, variously called *S-matrix theory* or *bootstrap theory*, that was expressed solely in terms of momenta and energy (specifically, four-momenta). This approach reflected the fact that the fundamental processes of physics are particle-interaction events and that experiments at accelerators are designed to measure the energies and components of momenta of the particles colliding with one another and their resulting products. The model proved to have little predictive power and was abandoned after the discovery of quarks, which were well described in traditional atomic fashion as localized particles moving in space and time. However, it should be noted that one of the ideas that arose in S-matrix theory, the *Veneziano model*, provided the initial mathematical motivation for what has developed into *string theory*. It remains to be seen if this variation, which involves not only four-dimensional space-time but also further spatial dimensions curled up on the Planck scale, will prove successful.

In our current, working four-dimensional space-time picture, particles are defined as objects that appear to be pointlike, that is, have no measurable spatial extension. Observations have led us to hypothesize that all objects with measurable spatial extension can be reduced to assemblies of elementary point particles. The current standard model of pointlike quarks, leptons, and gauge bosons agrees with all the data. Should the elementary objects turn out to be strings or *m-branes* (a point particle is a 0-brane, a string is a 1-brane, a sheet is a 2-brane, and so on), they would be so tiny that they are likely to continue to look like point particles for the foreseeable future.

SEEKING OBJECTIVITY

Given our existing, successful particulate model of reality, formulated within a space-time framework, we can begin to incorporate that model into the history of human observations. As we have seen, physicists have found that it does not suffice to simply describe those data from the point of view of a single observer, as we do when we take photographs of an object from different angles. Such models would not be wrong, in the sense that they would still correctly describe observations. However, they would be subjective and not likely to represent well what is "really" out there. Perhaps more important, these subjective models would be of no use to observers in other reference frames. Science always seeks objectivity because history has shown that this approach leads to more useful results. In order to have a decent chance of representing an objective reality, physicists formulate their models in such a way that they are the same for all observers. When they do so, they find that little more has to be postulated in order to derive the most important of what are traditionally termed the "laws of physics." Indeed, we only need to make up a few mathematical rules, as simple as possible, add a large dose of randomness, and we have about all we need to describe nature.

This can be likened to what our brains do when they take information from two eyes, each essentially a two-dimensional photograph, with an added dimension of energy (color), and use it to produce the concept of a three-dimensional object (plus color) that is then the same for all observers.

Normally we think of the laws of physics as constraints on the behavior of matter. Certainly the conservation laws are conventionally presented in that way. For example, we say that a ball tossed in the air will only go as high as its energy permits. However, a century ago Emmy

Noether proved that the great conservation laws of energy, linear momentum, and angular momentum are deeply connected to space-time symmetries. Within that framework, energy is the mathematical generator of a time translation. But, as we have seen, energy is also a quantity that can be measured. Our eyes provide a measure of energy with their ability to record color. However, let us consider the generic device that measures energy called a *calorimeter*.

A simple, though very inaccurate calorimeter can be made of an insulated container of water with a thermometer measuring the temperature of the water. A moving body is then directed into the water where it is brought to rest. The initial kinetic energy of the body is then operationally defined as the heat capacity of the water multiplied by the measured temperature increase. The assumption being made here is that the lost kinetic energy of the body is converted into heat energy so that the total energy of the system remains constant. Rest energy can also be converted to heat if a chemical or nuclear reaction takes place in the water. Other possible energy sinks, such as the production of sound, should also be considered if one wants to be very precise.

Suppose we have no calorimeter or other object of any kind to impede the motion of the body. Then we would expect that body to continue its motion with constant energy. Now, what would happen if energy were not conserved? Then a body moving through empty space might slow down and come to a halt, or it could speed up. Indeed, nothing would prevent the body from moving faster and faster. Alternatively, its measured mass might change or we might have some combination of effects.

In conventional physics, mass is the measure we use for the static inertia of a body. Momentum is a measure of dynamic inertia, which Newton identified as the "quantity of motion" of a body. Energy is a measure of the capacity of a body to do work, where work is the useful application of a force. All of these quantities are operationally defined in elementary physics textbooks.

For simplicity, let us assume the mass of a body is constant, so any energy change can be determined from a measured change in speed. Then, imagine making a video of such a body moving through empty space. In the case where energy is constant, if we later view the video we cannot determine the absolute time of any particular frame. That is, we cannot assign a unique time to each video frame that is distinguishable from the time assigned to any other frame. On the other hand, when the energy is not constant—when energy conservation is violated as determined by a measured change in speed—each frame can be labeled with a unique time. And so we see the intimate connection between energy conservation and time-translation invariance.

What is happening with these phenomena? The traditional view is that a special "law" called *conservation of energy*, handed down from above or built into the structure of the Universe, is acting to keep an isolated body moving with constant energy.

I have suggested an alternate view. Energy is a quantity we physicists have invented, defined operationally by what is measured with a calorimeter. In order for our observations to be independent of any particular point of view—that is, in this case, not dependant on when we happen to start our clock—that quantity must be independent of the time read off that clock.

Let me attempt to make this statement more precise: If we want to build a space-time model of events that describes observations in such as way that those descriptions do not depend on the point of view of a specific observer, then that model must contain a quantity called energy, defined as what you read on a calorimeter, that is conserved. Similarly, it must contain conserved linear momenta and angular momenta.

Now, it can be argued that Helmholtz discovered energy conservation in 1847, well before Noether developed her theorem. However, in retrospect we can see that although energy was not explicitly part of Newtonian mechanics originally, Newton's laws of motion are time-translation invariant and therefore inherently imply energy conservation. In fact, energy conservation is routinely derived from Newton's laws in freshman physics classes.

FALSIFIABLE!

Similar arguments hold for the other conservation principles. Linear and angular momenta, electric charge, baryon number, and several other quantities in physics that are defined operationally in terms of their measurements must be conserved in any space-time model that does not depend on a particular point of view. These conservation laws, then, are not laws at all, in the usual sense of the term. In fact, they are kinds of "nonlaws," or what I like to call "lawless laws." To single out a specific position or direction in space-time would require some causal action that we might interpret as a law. But what we see instead is the absence of any such action. The conservation principles are then forced upon us by our need to keep our models free of any reference to special points and directions—by our need to be objective if we hope to describe an objective reality.

Let me give a serious example where, if things had turned out different, we would have had to discard the principle of point-of-view invariance. In the late 1920s the energy spectra of beta rays from various beta-radioactive atomic nuclei were measured. The basic process believed to be taking place was the decay of a neutron in the nucleus, $n \rightarrow p + e$, to a proton, p, and electron, e, the beta ray being identified with the electron. In the rest frame of the neutron, linear momentum conservation requires that the proton and electron go off with equal and opposite momenta, whose magnitude is fixed by energy conservation, with some of the rest energy of the neutron being converted into kinetic energy of the decay products. Since its linear momentum is fixed, the electron's kinetic energy is fixed. However, this is not what is observed. Rather, a continuous spectrum of electron energies is seen.

The implication was that either energy is not conserved or another, invisible particle is being produced. The latter possibility also would explain the apparent violation of angular momentum conservation, since the three particles in the reaction have spin-1/2. The orbital angular momentum quantum number must be an integer, so there is no way to balance angular momentum in the three-particle reaction.

In 1930 Wolfgang Pauli proposed that the invisible particle was a neutral, spin-1/2 particle of very low or zero mass, which Enrico Fermi dubbed the *neutrino*. It took another twenty-five years before the existence of the neutrino was finally confirmed in the laboratory.

Now, suppose that the neutrino had never been found—that the reaction always took place with just the two particles being produced. This would have implied that energy conservation, or possibly linear momentum conservation (or both), was violated, along with angular momentum conservation. The implication would then have been that the laws we introduce to describe our observations would depend on when and where we make those observations. That would not have been wrong, but things did not turn out that way.

This also shows why what I have presented should not be confused with postmodernism. In postmodernism, "anything goes." That is not true for physics.

In short, the principle of point-of-view invariance, which we have used to provide the foundation for our models, is an eminently testable, falsifiable principle. So far, it has not been falsified.

THE CONCEPT OF FORCE

Let us next recall our concept of force. In conventional mechanics, conservation principles are supplemented by equations of motion and force laws that together describe the motion of a body. In quantum mechanics, those equations determine the average motion of an ensemble of bodies. In the common view, a force is an agent that acts to make some event happen, such as a body changing its state of motion. However, we have found that all known fundamental forces are quantities that physicists introduce into their equations to preserve the invariance of those equations for exchanges between different points of view.

The centrifugal and Coriolis forces were introduced in classical mechanics, so we could still apply Newton's laws of motion in rotating reference frames. Note that these forces do no exist in all reference frames. If you sit above Earth, watching it rotate below you, you do not need to introduce centrifugal or Coriolis forces to describe the motions of bodies you observe on Earth using Newton's laws of motion. But if you insist on describing those motions using Newton's laws from a fixed spot on Earth, then you need to introduce these forces into the model to preserve point-of-view invariance.

Einstein realized that the gravitational force is similarly the consequence of the differing perspective of various reference frames. An observer in a falling elevator in a uniform gravitational field does not experience any gravitational force. That observer's situation cannot be distinguished from one in which she is inside a closed capsule far out in space, distant from any astronomical body. On the other hand, an observer on Earth watching the elevator fall observes acceleration and, in order to maintain Newton's second law, imagines gravitational force acting on the elevator. Indeed, the gravitational force shows up mathematically in our equations when we transform Newton's second law to a reference frame in which the observer in the elevator is seen to be accelerating. But that force is just as fictional as the centrifugal and Coriolis forces.

What then "causes" a body to accelerate to the ground? As far as our classical description is concerned, in which we single out specific reference frames, we can call it the gravitational force. Or, if we use general relativity, which is independent of reference frame, we say the body travels something akin to the shortest path in a curved space called a *geodesic*. It depends on the model. In this case, general relativity is the superior model. But in many cases, different models give equally adequate but different, even contradictory, explanations for the same phenomenon. Then it is hard to see how to favor one over the other as the "true reality."

Indeed, motion itself is our own construction. Recall that we defined space and time in terms of clock measurements and built a model of bodies "moving in space." The laws of force are such as to assure the consistency of this construction. Again, the process is not arbitrary. While motion and force are our inventions, since they are just the ingredients of a contrived model, not any old model will do. It must accurately describe observations.

The space-time symmetries assumed in our models include Galilean invariance, in which all reference frames moving at constant velocity with respect to one another are equivalent. The specific Galilean transformation, which takes you from one such reference frame to another, applies only for relative speeds much less than the speed of light. The more general invariance under a Lorentz transformation applies for all relative speeds and can be derived as a consequence of rotational invariance in four-dimensional space-time, another form of point-of-view invariance. Once we have the Lorentz transformation and suitable definitions for momentum and energy that are consistent with Lorentz invariance, we have all of special relativity, including time dilation, Fitzgerald-Lorentz contraction, and $E = mc^2$. That is, special relativity is also forced upon us by our desire to describe objective reality.

As we saw, along with Lorentz invariance some weak form of Mach's principle that I have called the Leibniz-Mach principle comes into the development of general relativity. This principle also can be seen to follow from our attempt to describe nature objectively. A body by itself in the Universe cannot accelerate. There is nothing with respect to which it can accelerate. A second body is needed.

Einstein recognized that the energy-momentum tensor at the position of a given particle could be used to covariantly represent the effect of all the other bodies that may be present in the vicinity—indeed, in the remainder of the Universe. That tensor is simply a Lorentz-invariant extension of the concept of mass or energy density. Density depends on reference frame; the energy-momentum tensor, treated as a single mathematical object, does not. Tensors are a generalization of vectors. Recall that a vector is invariant to rotations and translations of coordinates systems, while the components of the vector are not invariant. Tensors are also invariant to translations and rotations in space-time, although their components are not. Transformation equations allow you to compute the components as you move from one reference frame to another.

In Einstein's model, another tensor, now called the Einstein tensor, is equated to the energy-momentum tensor with a proportionality constant that includes Newton's gravitational constant G. This tensor can be thought of as the gravitational field in general relativity. In Einstein's specific mathematical implementation of the model, which need not be the only way to achieve the same results, the Einstein tensor is then assumed to be connected as simply as possible to the geometry of space at the given point in space-time, as specified by the metric tensor, which is allowed to vary. With no further assumptions, the basic equation of general relativity, Einstein's equation, can then be derived.

CONSTANTS

What about the constant G? Where does it come from? Like the speed of light c and Planck's constant $\hbar$, G is one of the constants of nature whose value is still regarded as mysterious by some physicists and almost everybody else. However, we have already seen that c and $\hbar$ are arbitrary conversion factors, and the same is true for G. The "strength" of the gravitational force is not an adjustable parameter! The specific value of

G is set by the unit system you choose to use. You can make all physical calculations in *Planck units*, where $\hbar = c = G = 1$ (sometimes, $8\pi G = 1$).

Now, this does not mean that the strength of gravity relative to the other forces of nature is arbitrary. That strength is represented by a dimensionless parameter that is, by convention, proportional to the square of the proton mass. In principle, any mass could have been used. However, in all cases the relative strength of gravity would depend on some mass parameter that would allow it to vary in relation to the dimensionless electromagnetic strength and other force strengths. Mass is to gravity what electric charge is to electromagnetism. The dimensionless numbers that measure the relative strengths of the forces might someday be calculable from fundamental theory, but currently we must determine their values in experiments. At this time, we must still allow for the possibility that their values may be accidental—possibly the result of spontaneous symmetry breaking.

GAUGE SYMMETRY

Gauge symmetry, or gauge invariance, represents a generalization of space-time symmetries to the abstract space of the state vectors of quantum mechanics, or to any other function space that physicists use to mathematically describe the dependence of a physical system on its observables. In fact, the space-time symmetries can be subsumed in a single principle of gauge symmetry by simply including the space-time coordinates as part of the set of abstract coordinates being used to specify the state of the system. We can view what physicists call the "gauge" as equivalent to a particular theorist's reference frame. A gauge transformation is then analogous to the Lorentz transformation, or a simple rotation of coordinates, which takes you from one reference frame to another. Gauge symmetry is then a generalized form of point-of-view invariance.

We have seen how the generators of gauge transformations, viewed as rotations in state vector space, are associated with common observables. These are conserved when the system is invariant to the corresponding gauge transformation. Note that this is a generalization of Noether's theorem, which originally considered only continuous space-time symmetries. When expressed as operators, these observables yield the familiar commutation rules of quantum mechanics. We saw that Heisenberg's use of matrices to represent observables and Schrödinger's alternative use of differential operators, which Dirac generalized as operators in linear vector spaces, implemented Noether's proposals. Basic quantum notions, such as the uncertainty principle, then follow from the mathematics, that is, with no additional physical assumptions. In particular, the superposition principle, which accounts for quantum interference effects and quantum state entanglement, follows from gauge symmetry.

A gauge transformation that is applied at every space-time position is called a *global gauge transformation*. When it is allowed to change from point to point in space-time, it is called a *local gauge transformation*. The power of local gauge symmetry, the invariance to local gauge transformations, was a major discovery of twentieth-century physics.

The simplest example of a local gauge transformation occurs when the state vector space is represented by a complex number, called the *wave function*, which is a solution of the Schrödinger equation for nonrelativistic spinless particles. When an electrically charged particle is moving free of any forces, that equation is not invariant to a local gauge transformation. In order to make it so, we must introduce additional fields. These fields correspond precisely to the electric and magnetic potential fields of classical electromagnetic theory. Indeed, the local gauge transformation in this case relates directly to the gauge transformation of potentials in classical electrodynamics. The electric and magnetic fields defined by Maxwell's equations follow from the potentials. The same result is obtained for the Dirac equation, which describes a relativistic, spin-1/2, charged particle, and for the Klein-Gordon equation, which describes a relativistic, spin-0 particle. The equation used to describe a neutral spin-1 particle is locally gauge invariant, provided the mass of the particle is identically zero. Thus, the zero mass of the photon is derived. Charge conservation is also seen to follow from global gauge invariance.

The choice of gauge in state vector space is analogous to the choice of reference frame in space-time. As with gravity, the observed effects we associate with electromagnetism and describe in terms of particle acceleration are forced upon us by the requirement that our descriptions of the data be point-of-view invariant. Like gravity, the electric and magnetic forces are fictional, introduced to preserve point-of-view invariance.

Moving on to the other forces of nature, the standard model represents these as well in terms of gauge symmetries. Higher-dimensional state vector spaces are needed and this leads to a multiplicity of gauge bosons analogous to the photon. In the case of the weak force, three gauge bosons correspond to the three gauge fields, which were needed in view of the fact that, unlike electromagnetism, the weak force can involve the exchange of electric charge between interacting particles, which is not possible in electromagnetic interactions since the photon is electrically neutral.

Experiments over many years led physicists to realize that gauge symmetry was in fact broken in the observed weak force. The primary evidence was the extremely short range of the interaction, which implies that the weak gauge bosons have large masses. As mentioned, local gauge invariance requires that spin-1 particles be massless.

SPONTANEOUS SYMMETRY BREAKING

And so, a picture has developed in which symmetries are valid at very high energies, but at lower energies some symmetries are broken. This symmetry breaking could be dynamical, that is, as the result of some "lawful" higher process lying still undiscovered. However, in a simpler alternative, the symmetries are broken "spontaneously" by a phase transition analogous to the breaking of symmetry as a magnet cools below the Curie point. In the process, the fundamentally massless elementary gauge bosons gain mass by a process called the *Higgs mechanism*. Note that when the symmetry breaking is spontaneous, the underlying model remains symmetric.

In the case of the strong interaction, an even higher dimensional state vector space is needed, in which eight fields are introduced to preserve gauge symmetry with eight gluons comprising the field quanta. Gauge symmetry remains unbroken, so the gluons are massless. The finite range of the strong interaction is described as resulting from a color discharge that results when quarks are pulled too far apart.

Although gauge symmetry is maintained in two of the three sectors of the standard model, a completely symmetric model would be expected to occupy a single, unified sector. Despite a quarter century of attempts to find that underlying symmetry, its exact nature has not yet been uncovered. The problem is that there are many mathematical possibilities and only higher-energy data can settle the issue. Point-of-view invariance just requires that things be symmetrical, not that any particular mathematical symmetry group—which are all human inventions anyway—is uniquely specified. In fact, since the space-time-gauge model is our invention, perhaps many alternative models can be found. Indeed, one of the current problems with string theory is that almost an infinite number of possible models seem to exist.[1]

Some evidence and considerable theoretical predisposition supports the proposition that supersymmetry, the invariance under transformations between bosons and fermions that is not present at the "low" energies of current experiments, will be a major ingredient of any future unified model that comes in at sufficiently high energy.

Without speculating on the precise form future models may take, we can reasonably assume that they will reduce to the current standard model at the energies of current experimentation. These energies are already high enough to describe the physics into the heart of the big bang when the Universe was about a trillionth of a second old. This was well before the atoms and nuclei that make up the current visible Universe were formed. From that point forward in time, at least, we can say the Universe is comprehended. That is, we now have a model that describes the fundamental principles underlying all of current human experience and observation. We also have plausible, comprehensible scenarios consistent with all known physics that can account for a possible time before the big bang within a model of endless space and time with no underlying arrow of time.

This is the "comprehensible cosmos" of the title of this chapter and book. The fact that not all the parameters of the standard model are determined by that model but must be set by experiment is not a major problem. The number of parameters is far less than the totality of measurements that are described. The notion that we will someday reach the stage where everything is calculable from a few fundamental principles is probably nothing more than a pious hope. In the view presented in this book, the values of some of the parameters of physics may be accidental—the result of spontaneous symmetry breaking or some other indeterministic mechanism. Or they may simply be different in different universes and we happen to be fine-tuned to the parameters of the Universe in which we live.

Although we cannot calculate every parameter or predict every future happening, we can still claim to understand the Universe—at least to the point where the unknown is not an unfathomable mystery forever beyond our ken that must be assigned to some agency beyond nature. We cannot predict a roll of dice, but we still comprehend the phenomenon.

The standard model suggests that spontaneous symmetry breaking is the key to the structural complexity of the Universe, including the unspecified parameters of our models. At the extreme high temperatures of the early Universe, we would expect a high level of symmetry and, ultimately—at the Planck time—no structure at all. Symmetry breaking must be included in our model of the early Universe in order to produce the large observed excess of matter over antimatter, the mass difference between the proton and the neutron, a short-range rather than a long-range weak force, and generally particles with nonzero mass. Since higher energy is associated with smaller distances, we can imagine that deep inside matter the higher symmetries prevail, thus providing the underlying foundation of matter. What we observe, even with our most powerful scientific instruments, is still the colder, larger distance behavior of matter.

It may be asked, how do we reconcile symmetry breaking with point-of-view invariance? Note that symmetry breaking occurs at the comparatively low energies of current experimentation and the even lower energies of common human experience. This corresponds to a particular point of view being singled out, one that is not independent of the relative motion of the observer and the observed. We have to violate point-of-view invariance to some extent in our models if they are to be useful in the specific perspective of low-energy human observations. Remember the symmetry breaking is in our models, so we are the ones doing the symmetry breaking. However, once again, this is not to be interpreted as meaning that reality is all in our heads. Those models still must describe objectively observed facts.

In recent years it has become fashionable to speak of the different physical parameters of the Universe as if they were preselected to lead to the evolution of life on Earth and other planets. The fact that life as we know it depends on the exact values of these parameters, "fine-tuned" to high precision, is regarded as evidence that the Universe was designed with that purpose in mind. The notion is termed the *anthropic principle*, a misnomer since humanity is not singled out in any way.[2] Many physicists believe that when the ultimate "theory of everything" (TOE) is discovered, all the parameters that currently must be determined by experiment will be derived from some minimum set of principles. It seems highly improbable, however, that any purely natural set of principles would be so intimately connected to the biological structures that happened to evolve on our particular planet. A more likely scenario is that life as we know it evolved in response to the particular set of physical parameters of the Universe.[3]

If the symmetry breaking in the early Universe was in fact spontaneous, that is, not subject to any underlying dynamical mechanism, then the values of at least some of the parameters are accidental. Put another way, if we had an ensemble of universes, then the parameter values in our Universe arose from a random distribution—with no external, causal agent designating one particular set.

LAWLESS LAWS OF THE VOID

The data indicate no need for such a causal agent to provide for a set of rules that matter must obey. Our observable Universe, in fact, looks just as it would be expected to look in the absence of any such agent. The laws of physics are "ruleless rules" or "lawless laws" that arise not from any plan but from the very lack of a plan. They are the laws of the void.

Consider the decays of radioactive nuclei. Hundreds of examples have been studied with great precision since the beginning of the twentieth century and all these data indicate that these events are random and uncaused.

As discussed in chapter 6, the observation of the exponential decay curve for radioactive nuclei is evidence that the probability for decay is the same for all time intervals of the same duration, independent of time. An exponential rather than a flat distribution in decay times is observed simply because the number of nuclei decrease with time as they decay. In terms of the language we have developed in this book, the decay probability possesses time-translation symmetry.

Now, physicists call the observed decay curve the "exponential decay law." While we might imagine that some external agent acts to "cause" the decay of a nucleus to take place at a random time, the more parsimonious explanation is that it is simply random. If a causal process were in effect, then you would expect it to have some preferred time distribution. We have no more reason to assume that a decay is caused than we do that a fair coin is caused to fall head or tails with equal probability. A far simpler conclusion is that no invisible causal agency, natural or supernatural, was involved.

In this example, the exponential decay "law" follows from time-translation symmetry. This is a law of the void, which has no special point in time. As we have seen, all the global principles of physics can be termed lawless laws of the void.

WHY IS THERE SOMETHING, RATHER THAN NOTHING?

Now, you might ask, if the Universe has the global properties of the void, then why is it not a pure void? The answer may be that the void is less stable than a universe of matter.

We often find in physics that highly symmetric systems are not as stable as ones of lower symmetry, the states of broken symmetry. That is, the less symmetric state has a lower energy level, and a system will naturally tend to drop down to its lowest energy state. A pencil balanced on one end has rotational symmetry about the vertical axis, but is unstable and loses this symmetry when it topples over to a lower energy state (see fig. 5.5, p. 103).

The snowflake is another example that bears reviewing. We are accustomed to seeing snowflakes melt, but that is only because we live in an environment where the temperature is usually above freezing and energy is available in the environment to destroy the structure. Place a snowflake in a completely insulated vacuum, far out in space, and it will, in principle, last indefinitely.[4]

The void is highly symmetric, so we might expect it to drop to a lower energy state of lesser symmetry spontaneously. Calculations based on well-established models lend support to the notion that highly symmetric states are generally (though not always) unstable.

Nobel laureate physicist Frank Wilczek has written the following, which nicely sums up the picture I have been attempting to draw:

> Modern theories of the interactions among elementary particles suggest that the universe can exist in different phases that are analogous in a way to the liquid and solid phases of water. In the various phases the properties of matter are different; for example, a certain particle might be massless in one phase but massive in another. The laws of physics are more symmetrical in some phases than they are in others, just as liquid water is more symmetrical than ice, in which the crystal lattice distinguishes certain positions and directions in space.
>
> In these models the most symmetrical phase of the universe turns out to be unstable. One can speculate that the universe began in the most symmetrical state possible and that in such a state no matter existed. The second state had slightly less symmetry, but it was also lower in energy. Eventually a patch of the less symmetrical phase appeared and grew rapidly. The energy released by the transition found form in the creation of particles. This event might be identified with the big bang. The electrical neutrality of the universe of particles would then be guaranteed, because the universe lacking matter had been electrically neutral. The lack of rotation in the universe could be understood as being among the conditions most favorable for the phase change and the subsequent growth, with all that the growth implied, including the cosmic asymmetry between matter and antimatter. The answer to the ancient question "Why is there something rather than nothing?" would then be that "nothing" is unstable.[5]

Does this imply that we have explained how the Universe came from nothing (assuming that it did)? The meaning of the word *nothing* is a source of endless debate. How do you define nothing? What are the characteristics of nothing needed to define it? If it has any characteristics, any properties, then would it not be something? In his book *Why There Is Something Rather than Nothing,* philosopher Bede Rundle concludes, "[T]here has to be something."[6]

I have defined the void as what you get when you remove all the matter and energy. No physical quantities are measurable in the void. The void does not kick back when you kick it. If this void is not "nothing," then I do not know what is. But if the void is unstable, then we have "something" existing as another phase or state of nothing, the way ice and steam are different phases of water.

NOTES

1. Leonard Susskind, *The Cosmic Landscape: String Theory and the Illusion of Intelligent Design* (New York: Little, Brown, 2006).

2. John D. Barrow and Frank J. Tipler, *The Anthropic Cosmological Principle* (Oxford: Oxford University Press, 1986), p. 1086.

3. Victor J. Stenger, "Natural Explanations for the Anthropic Coincidences," *Philo* 3, no. 2 (2001): 50–67.

4. In practice, cosmic rays would eventually tear it apart since we could never completely isolate the snowflake from them.

5. Frank Wilczek, "The Cosmice Asymmetry between Matter and Antimatter," *Scientific American* 243, no. 6 (1980): 82–90.

6. Bede Rundle, *Why There Is Something Rather than Nothing* (Oxford: Clarendon, 2004), p. ix.

CHAPTER TEN

MODELS OF REALITY

DOCTRINES

The philosophical doctrine called *operationalism* holds that the meanings of scientific terms and concepts are wholly determined by the procedures by which measurements are made.[1] We have seen that time, distance, velocity, energy, temperature, and all the other observables we use in physics are defined operationally. The other quantities and mathematical functions used in our models, which are not directly measured, are formed from previous operationally defined observables. The concepts we include in the models, such as a framework of space-time and motion, are convenient pictures based on observations that are used to describe and predict further observations.

However, operationalism so defined does not properly describe either the practice of physics or the view of most physicists, nor does it represent the view presented in this book, which I maintain is *realism*. While all our physical observables are indeed operationally defined, they are not wholly determined by our measuring procedures. Surely they have some connection to the real world, whatever that may be. Physics is not fantasy. Energy is not conserved in dreams or Roadrunner cartoons.

Most physicists are realists who hold that objects exist independent of our perception of them. Our models seek to describe an objective reality, and this prerequisite forms the basis of the primary principle from which we have seen the special mathematical concepts and principles of physics arise—point-of-view invariance.

Another philosophical term that might more accurately apply to the attitude adopted here is *instrumentalism*. This is the belief that statements or models may be used as tools for useful prediction without reference to their possible truth or falsity. Although physicists generally talk as if the elements of their models exist in one-to-one correspondence with elements of reality, the success of those models does not depend on any such metaphysical assumption. Certainly the models of physics contain some element of truth if they accurately describe the data. But describing the data is the only legitimate claim they can make and is the only fair way they should be evaluated. We should not label them "false" if they meet that criterion, and thus instrumentalism, as defined above, is not an accurate description of the approach of this book.

Most physicists tend to attribute a direct reality to the concepts, such as fields and particles, that form the ingredients of their models. However, from my experience in talking with them and listening to their lectures for over four decades, I think that they are not so much staking out a specific metaphysical position as much as simply viewing metaphysics as a subject that cannot be adjudicated empirically, and hence not a subject that scientists can usefully contemplate. Scientific criteria cannot distinguish between viable metaphysical schemes. But they still can indicate which schemes are viable. In that regard, scientific consideration of metaphysics is not without value.

ATOMS AND THE VOID

The success of physics testifies to some connection to an underlying, objective reality. So what might that reality be? What metaphysical conclusions might we draw from physical models?

We have defined as matter anything that kicks back when you kick it. We call the things that kick back "objects" or "bodies." The success of the standard model of particles and forces suggests that the elementary objects of reality are the particles of that model, or perhaps some more elementary constituents yet to be uncovered. Thus, we can justifiably picture reality as envisaged by Democritus—atoms moving around in an otherwise empty void. We can call this metaphysical model *atoms and the void*.[2]

In this regard, the term *atom* is being used in its original sense to apply to the ultimate objects that cannot be divided into more elementary constituents. It derives from the Greek *atomos*, which means "that which cannot be cut into smaller pieces." Of course, now we know that the "atoms" of the nineteenth century, which were identified with the chemical elements, can be broken apart into electrons and nuclei. Atomic nuclei contain protons and neutrons, and these in turn are composed of quarks. In the standard model, electrons and quarks are elementary. If someday these are found to be composed of more elementary objects, such as strings or *m*-branes, then the basic metaphysical picture of atoms and the void remains unchanged except for the precise identity of the atoms.

We have seen that space, time, and motion are part of a systematic description of observations that provides no independent evidence to confirm their objective existence. Space or the void does not kick back. Motion does not kick back. All that kick back are, by definition, material bodies. The information they kick back to us leads us to assign to these bodies intrinsic properties such as mass, electric charge, spin, isospin, baryon number, lepton number, and so on, which are all operationally defined. That is, they are empirical but not wholly independent of our theoretical constructs. Within the space-time model, rules of behavior for these properties, called *laws*, arise from applying point-of-view invariance in a particular, idiosyncratic mathematical scheme that might have been formulated in a number of completely different ways if physicists in different cultures had been isolated from one another. Again, this does not imply any old scheme will do, or that physics is just another cultural narrative. The scheme still has to work. The particular scheme adopted by the world's physicists does work, but this does not allow us to conclude that it is the only one possible.

As it is, physicists form a global community that speaks the common language of mathematics, which has led them to adopt a common set of physical models. Within these models of humanity's mutual contrivance, observables are mathematically represented as generators of the various transformations we can make in the abstract space we use to specify the physical state of a system, where space-time is now included as part of that abstract space.

The case of particle spin exemplifies the rather subtle point being made. This was discussed in chapter 4, but let us recall the argument. In classical physics, a rotating body has a total angular momentum about its axis called *spin*. A point particle necessarily has zero-spin angular momentum, because it has no subparticles circulating about its center. In quantum physics, objects such as electrons, which we treat as point particles, nevertheless possess spin. This is measured by the interaction of the particle's magnetic field with an external magnetic field. However, simply assuming that a quantum particle is really a spinning body of finite extent does not give the correct quantitative answer. In particular, the spin "quantum number" so conceived can only be an integer or zero, since the total angular momentum of the body results from summing the orbital angular momenta of all the particles inside the body. Orbital angular momentum quantum numbers can only be zero or integers. Half-integer-spin quantum numbers do not arise in this picture.

The view developed in this book gives an explanation for particle spin that includes the possibility of half-integers. Assume a model of pointlike particles localized in three-dimensional space. A point possesses rotation symmetry in space. The generators of rotations about each axis are the angular momentum component operators and these correspond to conserved quantities that can be operationally defined and measured. The multiplicative rules for these operators can be used to prove that the total angular momentum quantum number can be zero, integer, *or half-integer*.

So, we conclude that if we are to utilize a model to describe phenomena that assumes three-dimensional space and postulates pointlike particles within that space, then point-of-view invariance forces us to include as part of that model spins that are zero, integer, or half-integer. This does not require that spinning point particles truly exist in the underlying reality—just that our specific model of such particles requires that they have the abstract property of spin.

Similarly, the other space-time symmetries require that a particle must be described as having mass, energy, and linear momentum. They must obey the rules of special and general relativity. Gauge invariance will require that the particles have electric charge (which can be zero), isotopic spin, and other quantities depending on the gauge transformation group being used in the model. If the particle has spin-1, it must be massless unless local gauge invariance is broken.

Once again, these are not necessarily real properties of real particles. They are the properties that the particles in our model space-time must have if that model is to describe observations objectively. They are artifacts of our special choice of models. They might not occur in some other model.

Furthermore, mathematical quantities do not kick back when you kick them, except in high school math courses. Nevertheless, we can postulate a metaphysical model in which the primitive ingredients are bodies that possess objective, measurable properties, which we interpret as mass, electric change, spin, isospin, baryon number, lepton number, and so on. In that model, atoms and the void, quarks and electrons are real.

ALTERNATE REALITIES

At least, this is one possible metaphysical view. Other possibilities cannot be ruled out. Indeed some, if not most, mathematicians and theoretical physicists hold an opposing view in which our most abstract mathematical notions, such as wave functions, quantum fields, metric tensors, and the very equations that connect them all together, exist in another world separate from the physical world—a world of ideal, Platonic "forms."

This notion has been fully expounded in a monumental new work by Oxford mathematician Roger Penrose titled *The Road to Reality: A Complete Guide to the Laws of the Universe*.[3] As in his earlier work, *The Emperor's New Mind*,[4] Penrose argues that mathematical truth exists in some kind of external, objective reality of Platonic forms that is not simply a human invention.

Penrose explains what he means by the "existence" of a Platonic realm of mathematical forms:

> What I mean by this "existence" is really just the objectivity of mathematical truth. Platonic existence, as I see it, refers to the existence of an objective external standard that is not dependent upon our individual opinions nor upon our particular culture. Such "existence" could also refer to things other than mathematics, such as to morality or aesthetics, but I am here concerned with mathematical objectivity, which seems to be a much clearer issue.[5]

Penrose divides "reality" into three "worlds": physical, mental, and mathematical. While he does not specifically define the physical world, we can clearly associate it with my definition of material bodies, namely, that which kicks back when you kick it. The mental world is that subset of the physical world that includes the phenomena we associate with thoughts.

Penrose's mathematical world is the set of all "true" statements, including those that have nothing do to with the physical world and so is not wholly part of that world (or its subset, the mental world). In this regard, by "true" Penrose means logically consistent as, for example, a provable mathematical theorem. According to Penrose, the Platonic mathematical world is separate from the physical world and the mental world. Only a small part of the mathematical world encompasses the physical world. Much of "pure" mathematics is not motivated by observations of the world. Similarly, only a small part of the physical world induces mentality (rocks don't think) and only a small part of the mental world is concerned with absolute mathematical truth. However, he does seem to think that the mental world is fully induced by the physical world. He maintains his previous position in *The Emperor's New Mind*, where he argued that mental processes are not purely computational, as with a computer, but that they are still physical. He probably can even do away with the mental world in his current scheme, incorporating it as part of the physical world (as I have above).

Some philosophers of mathematics regard the mathematical world as purely the product of the mental world. Penrose disagrees:

> The precision, reliability, and consistency that are required by our scientific theories demand something beyond any one of our individual (untrustworthy) minds. In mathematics we find a far greater robustness than can be located in any particular mind. Does this not point to something outside ourselves with a reality that lies beyond what each individual can achieve?[6]

And,

> Mathematics itself indeed seems to have a robustness that goes beyond what any mathematician is capable of perceiving. Those who work in this subject . . . usually feel that they are merely explorers in a world that lies beyond themselves—a world which possesses an objectivity that transcends mere opinion.[7]

On the other hand, Penrose regards the physical world as fully depicted by mathematical laws. This is not inconsistent with the position I have presented in this book. If the principles of physics belong to a separate, mathematical world, then they are not part of the physical world! As I have indicated, the models of physics represent idealized mathematical representations of observational data, formulated to be independent of individual viewpoints and culture—just like the mathematics of which Penrose speaks. Pointlike particles, wave functions, and indeed space and time themselves may be thought of as living in the Platonic world of mathematical forms, a world separate from the physical world. The existence of such a separate world was suggested to Penrose by the apparent objective nature of mathematical truth. Similarly, the fact that physical models based on point-of-view invariance agree so well with objective observations suggests that they may also carry a reality that transcends the physical and mental worlds.

Still, we cannot demonstrate the existence of a metaphysical Platonic world by logic or data. If we could, it would be physics and not metaphysics. And we cannot prove atoms and the void, in which Penrose's mathematical and metal worlds are subsumed. All we can argue is that the model of atoms and the void possesses a natural simplicity that the reader might find easier to comprehend. At the very least it provides a simple picture of a comprehensible cosmos that is consistent with all we know from observation and logical analysis.

The unsettled dispute over alternate metaphysical models arises out of quantum mechanics. Atoms were pretty easy to accept in the prequantum, Newtonian world as long as they were limited to their association with matter and were not proposed as an alternative to the theological concept of the immortal, immaterial soul. Although direct evidence for atoms was not obtained and their structures not elucidated until early in the twentieth century, atoms became a basic part of physical models by the end of the nineteenth century.

ATOMS AND THE ETHER

Still, prior to the twentieth century atoms alone did not appear sufficient to explain all natural phenomena; some, such as light and gravity, did not appear atomic. Newton originally proposed an atomic or "corpuscular" model of light and speculated about gravity also resulting from the flow of particles. However, these phenomena seemed to require additional ingredients besides localized particles swimming about in an empty void. Those added ingredients were called *fields*. In the nineteenth century, Newtonian gravity and Maxwellian electromagnetism, which included the model of light, were described as continuous field phenomena—tensions and vibrations of a smooth, elastic medium called the *ether* that was postulated to pervade all space. Indeed, a reasonable metaphysical scheme based on knowledge at that time could have been termed *atoms and the ether*, with the ether forming the "substance" of space and a totally empty void nonexistent, as Aristotle had argued two millennia earlier.

However, when late in the nineteenth century Michelson and Morley kicked the ether, it did not kick back. A few years later Einstein did away with the ether and his photon model of light replaced ethereal vibrations with particles of light. Along with the strong empirical support for the atomic model of matter, light was found to be a discrete, localized phenomenon.

Still, Democritus was not fully vindicated. The wavelike behavior of light did not go away, and particles, such as electrons, were also found to exhibit phenomena, such as interference, which occur for waves in familiar materials like water and air. This "wave-particle dualism" has led to a century of disputation over what quantum mechanics "really means" that is still not settled.

The common refrain that is heard in elementary discussions of quantum mechanics is that a physical object is in some sense both a wave and a particle, with its wave nature apparent when you measure a wave property such as wavelength, and its particle nature apparent when you measure a particle property such as position.

But this is, at best, misleading and, at worst, wrong. Whenever you detect photons, electrons, or other elementary objects, you are detecting localizable phenomena. For example, when you send a beam of monochromatic light, say, from a laser, through a tiny slit and then let the light passing through the slit impinge on an array of small photodetectors sensitive at the one-photon level, you will get individual, localized hits for each photon. You can do the same for a monoenergetic beam of electrons and an array of electron detectors. When you plot the statistical distribution of hits, you will see the familiar diffraction pattern indicative of waves. From measurements of this pattern, you can obtain an effective "wavelength."

That wavelength will be exactly equal to Planck's constant divided by the momentum of the particle, as proposed by Louis de Broglie in the 1920s. In other words, measuring the wavelength, a wave property, is equivalent to measuring the momentum, normally thought of as a particle property.[8] Furthermore, wave effects are seen in the probability distribution of an ensemble of particles and not a single object, particle or wave.

Indeed, the concept of waves need never be introduced in fundamental physics—a point that was made by both Dirac and Feynman. In Dirac's classic *The Principles of Quantum Mechanics*, the term *wave function* is mentioned just once—in a dismissive footnote: "The reason for this name [wave function] is that in the early days of quantum mechanics all the examples of those functions were of the form of waves. The name is not a descriptive one from the point of view of the modern general theory."[9]

Feynman was similarly dismissive of waves. In his series of lectures before high school students that was published in a small book called *QED*, he remarked: "I want to emphasize that light comes in this form—particles. It is very important to know that light behaves like particles, especially for those of you who have gone to school, where you were probably told something about light behaving like waves. I'm telling you the way it does behave—like particles."[10]

Quantum mechanics, in its conventional application, treats the interference pattern as a probability distribution of an ensemble of particles, not as the predictive behavior of a single entity called a "wave." No etheric medium is doing any waving. The wave function lives in a multidimensional abstract space, not familiar space-time. So it is hardly a likely candidate for a component of reality unless you take the Platonic view.

Our most advanced way of describing fundamental physical phenomena is in terms of relativistic quantum field theory. An oscillating, propagating field is mathematically a wave, that is, it obeys a wave equation. In one possible metaphysical model, these fields can be taken as the "true reality." However, in quantum field theory, for every field there exists an associated particle called the *quantum* of the field. Indeed, all particles are viewed as the quanta of fields. The success of quantum field theory then does not demand a reality of waves or fields. The particle quanta can be alternatively regarded as real. If waves are real, then what is the medium that's doing the waving? It seems that some kind of ether then has to be reintroduced into physics and we have to then specify what composes this new ether. Furthermore, this ether lives in some multidimensional, abstract space—not familiar space-time.

PARTICLE REALITY

The metaphysical model of atoms and the void has the virtue of simplicity and ease of comprehension. In this scheme, fundamental particles (or perhaps strings or *m*-branes) and the more complex objects that are formed when they stick together constitute the sole reality. The observed behavior of these objects is described in the very familiar way we describe the motion of rocks, balls, planets, and other localized bodies of everyday experience—with some differences needed to account for quantum phenomena. Physical bodies are pictured as moving through space and time in all directions, including backward, with respect to our familiar time direction. They interact by colliding with one another or exchanging other particles that carry energy, momentum, and other quantities from one point in space to another. Everything happens "locally." Even if two interacting particles are separated from each other in space as viewed from a particular reference frame, that separation is "timelike." We can always find a second reference frame moving no faster than the speed of light with respect to the first in which the two events are at the same position; that is, they are "local," and so they differ only in their time coordinates.

By "event" I mean a phenomenon that occurs at a certain point (or localized region) in space and time. For example, an event **A** might be a nucleus emitting an electron. Event **B** might be an electron being absorbed by another nucleus. The two events can be indicated as points in a space-time diagram, as illustrated in fig. 10.1.

In the case shown, event **B** lies outside the "light cone" of event **A**. This (in general, four-dimensional) cone specifies the region in space-time that includes all the events that can be reached by a signal from **A** traveling at the speed of light or less. The separation of events outside the light cone is said to be "spacelike," or *nonlocal*. It is not possible to find a second reference frame where the two events are at the same place,

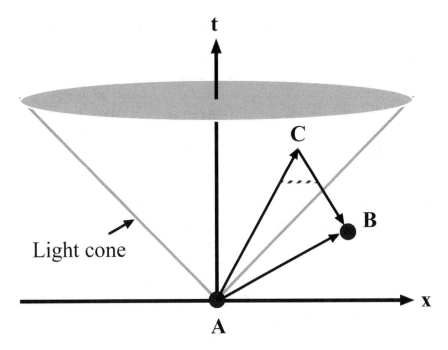

Fig. 10.1. Event **B** lies outside the light cone of event **A**. A signal directly from **A** to **B** must go faster than the speed of light. However, a signal moving at less than the speed of light can reach from **A** to **B** by going first to point C and then turning around and going backward in time to **B**. Thus, events outside their respective light cones can be connected without invoking superluminality (nonlocality), provided signals can propagate in either time direction. Nothing in physics prevents this from happening. Note, however, that if the signal from **A** to **B** is carried by a single particle, that particle appears at two places simultaneously, for example, as indicated by the dashed line.

unless it moves faster than the speed of light with respect to the first. From what we know about electrons, they travel at less than the speed of light, so the electron emitted from the nucleus at **A** cannot reach the nucleus at **B**.

The fact that **B** is outside the light cone of **A** means that any direct signal between the two must travel faster than the speed of light. This presents no problem for macroscopic phenomena, where superluminal speeds are never observed. The two events are simply not causally connected. They can have no possible influence on each other, and all observations confirm this. Note that another way to view this situation is that no reference frame exists where the two events are at the same place in space.

At the quantum level, some phenomena are thought to be nonlocal. That is, connections seem to exist between quantum events that lie outside each other's light cone, including the type of process described above. This implies, at first glance, that a signal connecting the two events would move faster than the speed of light.

The simplest example is two-slit interference. The observed interference pattern implies that a photon passing through one slit somehow communicates with a photon passing simultaneously through the other slit, which requires superluminal motion. Einstein called this "spooky action at a distance."

Many physicists believe that some form of nonlocal connections in quantum events have been confirmed by the experimental violations of Bell's theorem in so-called EPR (Einstein-Podolsky-Rosen) type experiments.[11] However, no actual superluminal signals have ever been recorded and are proscribed by the special theory of relativity.[12] Furthermore, standard quantum mechanics, which does not violate special relativity, yields the precise results observed in the EPR experiments, not to mention the interference patterns observed and many other phenomena that defy classical explanation. This has led to thousands of papers attempting to solve the apparent paradox between the supposed nonlocality of quantum events and the total absence of either observed or theoretical violations of Einstein's speed limit.

In addition, numerous popular books and articles exploit the "spookiness" of quantum mechanics to justify spooky paranormal claims, from *extrasensory perception* and mind over matter to various nonscientific healing schemes. I debunked these claims in my 1995 book *The Unconscious Quantum*.[13]

The EPR experiments and their interpretations are discussed in both *The Unconscious Quantum* and my 2000 book *Timeless Reality*.[14] Treatments also can be found in many other books, so I will not repeat the details here. Rather, I will show that a simple solution to the nonlocality problem exists that is consistent with all we know from the data.

As we saw in fig. 10.1, event **B** lies outside the light cone of event **A**.

A direct signal from **A** to **B** must travel at greater than the speed of light. However, a signal can also go from **A** to **B** by first going to **C** and then to **B**. Note that **C** is within the light cones of **A** and **B**. If the signal turns around at **C** and goes in the *opposite* time direction, it can reach **B**, again without exceeding the speed of light. Thus, superluminality need not be invoked if we allow propagation in both time directions.

I have emphasized several places in this book that no direction of time is evident in any fundamental observations. Nothing prevents a single molecule of outside air from finding its way inside a punctured, deflated tire. In principle, the tire can reinflate if sufficient molecules are pointing in the right direction. That this does not happen in experience is a matter of probabilities, not fundamental physics. The direction of time of our experience can be understood as a statistical preference in which many body systems, with their highly random behavior, tend to approach equilibrium. The second law of thermodynamics simply defines the arrow of time.

In 1953 Olivier Costa de Beauregard showed how to describe the results of EPR experiments by allowing particles to move in either time direction.[15] In general, time reversibility can be used to eliminate many of the so-called paradoxes of quantum mechanics. One easy way to see this that can be applied to any experiment is to imagine filming the experiment and running the film backward. If what you see does not strike you as spooky, then no reason exists to regard the experiment viewed in conventional "forward" time as spooky.

In 1949 Feynman introduced the notion that positrons can be viewed as electrons going backward in time, which beautifully explained their properties (see fig. 5.4, p. 101).[16] However, along with the majority of the physics community, he preferred to stick with the convention of a single time direction and interpret those electrons going in an opposite time direction as antiparticles. Nevertheless, this view is empirically indistinguishable from the one I propose, which is also more parsimonious in halving the number of fundamental objects. Directional time requires an additional hypothesis, not required by the data and not part of mechanics, either classical or quantum. The model of localized atoms and the void can be made consistent with all observations to date, including quantum phenomena, with a minimum of assumptions once we allow for time reversibility.

Note that if the signal from **A** to **C** to **B** in fig. 10.1 is carried by a single particle, then that particle appears at two places simultaneously, as indicated by the dashed line. Indeed, a particle zigzagging throughout space-time can appear many places at once. This is a kind of nonlocality that I prefer to call *multilocality* since the particle never needs to travel faster than c. Every point along a zigzag path in space-time is a local event.

The usual objection that is raised against motion backward in conventional time is the so-called grandfather paradox. If one could travel back in time, he could kill his own grandfather. However, it can be shown that this paradox does not hold at the quantum level, where all grandfathers are indistinguishable.[17]

In *Timeless Reality* I showed that time reversal makes it possible to reify many of the features of modern quantum and particle physics. For example, the multiple particle paths in the Feynman path integral formulation of quantum mechanics[18] can be pictured as actually occurring in a single event. Thus problems like multiple-slit interference and diffraction, which are treated in the Feynman formalism as if all possible paths occur, can be imagined as actually occurring as the particle zigzags in space-time.

Similarly, the multiple Feynman diagrams used to describe a single fundamental particle interaction can be viewed as all taking place in the event.[18] Time reversal can also be used to show how the so-called virtual particles in these diagrams, which are normally thought of as having imaginary mass (and thus, in principle, should travel faster than light!), can be viewed as real particles with normal mass moving at speeds less than or equal to c.

WHY DOES IT WORK?

So, I believe I have shown that a purely local particulate model of reality is viable. Indeed, it might even be a correct depiction of the fundamental structure of the world, just as real as planets and mountains. It is certainly a simple picture that we can readily use in our attempt to comprehend the cosmos.

However, I must again emphasize that none of the main results of this book in any way depend on this or any assumed metaphysics, which makes those results considerably stronger since they do not require me to demonstrate the truth of that metaphysics—an impossible task. My results derive purely and simply from applying the rule that our model be point-of-view invariant with respect to observations and the need to agree with those observations.

So, where does point-of-view invariance come from? It comes simply from the apparent existence of an objective reality—independent of its detailed structure. Indeed, the success of point-of-view invariance can be said to provide evidence for the existence of an objective reality. Our dreams are not point-of-view invariant. If the Universe were all in our heads, our models would not be point-of-view invariant.

Point-of-view invariance generally is used to predict what an observer in a second reference frame will measure given the measurements made in the first reference frame. For example, if I measure the time and distance interval between two events in my reference frame, then I can use the Lorentz transformation to predict what you will measure in your reference frame, provided I know your velocity relative to me and provided that it is constant.

Of course, a trivial application of point-of-view invariance occurs for multiple observers in the same reference frame. They should all obtain the same measurements within measurement errors.

Point-of-view invariance is thus the mechanism by which we enforce objectivity. If we did not have an underlying objective reality, then we would not expect to be able to describe observations in a way that is independent of reference frame.

While the reader may be disappointed that I have not taken a stronger stance on the actual details of the nature of reality, I believe that my cautious approach is more fruitful and, indeed, more realistic. I have emphasized that our models are contrivances, like cars. They have a job to do, namely, describe the data. When an old car doesn't work anymore, we buy a new one. Similarly, we can always replace an old model with a new one, without viewing it as some great revolution or paradigm shift. As my friend Edward Weinmann put it in an e-mail,

> One's world is not destroyed when an old model is discarded, and one is not a made a fool of by discoveries of anomalies. They are welcomed,

not dreaded. They are elements of good fortune, enabling us—even forcing us —to progress in knowledge. This would be the basis of the faith of a modern humanity—that we cannot persist in any current error, but will emerge from it, guided not by any supernatural force, but by adherence to modern scientific method. Of course this does not imply that there need be any end to this process, and that is a major departure from the attitude of centuries past.[20]

Perhaps we will never comprehend the true structure of reality. And perhaps we should not care. We have a model that fits the data, which is all we need for any conceivable application, save philosophical or theological discourse. We are prepared to modify that model as the data require. So we can leave it at that.

NOTES

1. The philosophical terms used in this section have been taken from the Philosophical Dictionary at http://www.philosophypages.com.

2. Victor J. Stenger, *Timeless Reality: Symmetry, Simplicity, and Multiple Universes* (Amherst, NY: Prometheus Books, 2000).

3. Roger Penrose, *The Road to Reality: A Complete Guide to the Laws of the Universe* (New York: Knopf, 2004).

4. Roger Penrose, *The Emperor's New Mind: Concerning Computers, Minds, and the Laws of Physics* (Oxford: Oxford University Press, 1989).

5. Penrose, *The Road to Reality*, p. 13.

6. Ibid., p. 12.

7. Ibid., p. 13.

8. Although waves in classical physics do carry momentum.

9. P. A. M. Dirac, *The Principles of Quantum Mechanics* (Oxford: Oxford University Press, 1930). This book has had four editions and at least twelve separate printings. The quotation here can be found in the footnote on page 80 of the 1989 paperback edition.

10. Richard P. Feynman, *QED: The Strange Theory of Light and Matter* (Princeton, NJ: Princeton University Press, 1985), p. 15.

11. A. Einstein, B. Podolsky, and N. Rosen. "Can the Quantum Mechanical Description of Physical Reality Be Considered Complete?" *Physical Review* 47 (1935): 777–80; John S. Bell, "On the Einstein-Podolsky-Rosen Paradox," *Physics* 1 (1964): 195–200; Alain Aspect, Phillipe Grangier, and Roger Gerard,

"Experimental Realization of the Einstein-Podolsky-Rosen Gedankenexperiment: A New Violation of Bell's Inequalities," *Physical Review Letters* 49 (1982): 91–94; "Experimental Tests of Bell's Inequalities Using Time-Varying Analyzers," *Physical Review Letters* 49 (1982): 1804–1809.

12. Technically, special relativity allows for particles, called *tachyons*, that travel faster than light. However, they must *always* travel faster. The theory forbids particles traveling at less than the speed of light from being accelerated past the speed of light. No tachyons have ever been observed.

13. Victor J. Stenger, *The Unconcious Quantum: Metaphysics in Modern Physics and Cosmology* (Amherst, NY: Prometheus Books, 1995).

14. Stenger, *Timeless Reality*.

15. Olivier Costa de Beauregard, "Une réponse à l'argument dirigé par Einstein, Podolsky, et Rosen contre l'interpretation bohrienne de phénomènes quantiques," *Comptes Rendus* 236 (1953): 1632–34.

16. Richard P. Feynman, "The Theory of Positrons," *Physical Review* 76 (1949): 749–59.

17. Stenger, *Timeless Reality*, pp. 203–208.

18. Richard P. Feynman and A. R. Hibbs, *Quantum Mechanics and Path Integrals* (New York: McGraw-Hill, 1965).

19. Richard P. Feynman, "Space-Time Approach to Non-relativistic Quantum Mechanics," *Reviews of Modern Physics* 20 (1948): 367–87.

20. Edward Weinmann, e-mail message to author.

MATHEMATICAL SUPPLEMENT A

THE SPACE-TIME MODEL

PHYSICAL QUANTITIES

All physics quantities reduce to measurements made on clocks or constructs of those measurements. Since the only arbitrary unit is the unit of time, all measuring instruments must be ultimately calibrated, at last in principle, against the standard clock. Furthermore, every measurement unit in physics can be expressed as s^n, where s stands for *second* and n is a positive or negative integer, or zero. Some examples are given in the following table.

time, distance	s^1
velocity, angular momentum, electric charge	s^0
acceleration, mass, energy, momentum, temperature	s^{-1}
force, electric and magnetic fields	s^{-2}
Newton's constant G	s^2
pressure, mass density, energy density	s^{-4}

MODELING SPACE AND TIME

Consider two observers, Alf and Beth, as shown in fig. 2.1 (p. 40). (Note that figure references in the mathematical supplements will often point back to the main text. Figures specific to the supplements will contain identifying letters.) Each can emit a sequence of electromagnetic pulses with a fixed time interval or *period*, T_1. They detect any electromagnetic

190

pulses sent to them and measure the periods if the received pulses T_2 along with the time interval Δt between the first pulse transmitted and the first returned.

If $T_1 = T_2$ they are defined to be "at rest" with respect to each other. Suppose Alf measures $\Delta t > 0$, as seen in (b). He assumes that "out there" is an object, Beth, who reflected the pulses back to him. He imagines that a dimension called *space* separates Beth from him. He pictures the pulse as "traveling through space," taking a time $\Delta t/2$ to reach Beth and another $\Delta t/2$ to get back. The two are thus viewed in the model as being separated from each other by a "distance"

$$d = c\frac{\Delta t}{2} \tag{A1}$$

where c is an arbitrary constant introduced to change the units of time from nanoseconds to meters or some other familiar unit of distance. Indeed, they can chose $c = 1$ and measure the distance in *light-nanoseconds* (about a foot). They call c the *speed* of the pulses, or the *speed of light*.

Now suppose that, on another occasion, Alf observes that the pulses reflected by Beth are received with a different period, $T_2 \neq T_1$ than those he transmitted, as in (c). He models this by hypothesizing that she is "moving" toward or away from him along their common line of sight. Alf can use the two periods to measure their relative velocity along that direction. As illustrated in the space-time diagram given in fig. A2.1,

$$vT_1 = c\left(T_2 - T_1\right) \tag{A2}$$

and Beth's velocity is

$$v = c\left(\frac{T_2}{T_1} - 1\right) \tag{A3}$$

Note that v will be positive when $T_2 > T_1$ and negative when $T_2 < T_1$.

ADDING DIMENSIONS

The preceding discussion applied for one dimension. After considerable experimentation with multiple observers, we find that we need a space of more than a single dimension in our model in order to agree with the data. Let us first add another observer, Cal, and assume that all three observers—Alf, Beth, and Cal —can communicate with one another by means of a suitable modulation of pulses or other signals. Then all three know the distances between one another, d_1, d_2, and d_3. We can represent these as a triangle shown in fig. 2.2 (p. 42).

This allows us to define another measure of space called *direction*, which is specified by a quantity called *angle*. The angle θ_1 between the direction from Alf to Beth and the direction from Alf to Cal is given by

$$\cos\theta_1 = \frac{d_1^2 + d_2^2 - d_3^2}{2d_1 d_2} \qquad (A4)$$

which is the familiar *law of cosines*. We can obtain the other two angles of the triangle similarly.

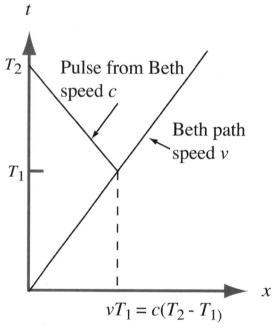

Fig. A2.1. Space-time diagram showing Beth's motion at a speed v in Alf's reference frame. Beth travels the distance vT_1 in the time T_1 since the last pulse was sent. The light pulse from Beth to Alf arrives at a time T_2. So Alf observes a returned pulse period T_2.

Note that this assumes a flat, Euclidean space. The three angles will add up to 180 degrees. In general relativity, where a non-Euclidean space is used, another procedure must be implemented. We can introduce a coordinate system by letting *x*-axis lie along Alf-Beth and draw a *y*-axis at right angles, as indicated by the dashed lines in fig. 2.2. These lines can also be defined by the paths of light rays and could, in principle, be "straight" or "curved."

Using the direction from Alf to Beth as a baseline, we can define the position displacement of Cal from Alf as the vector $\mathbf{d_2}$ whose *magnitude* is the single number d_2 and whose direction is the angle θ_1. A two-dimensional *x*, *y* coordinate system can be used to specify the positions. Then the vector displacement from Alf to Cal is $\mathbf{d_2} = (d_2, \theta_1) = (x, y) = (d_2\cos\theta_1, d_2\sin\theta_1)$.

If we add a fourth observer, Del, we find, by experimentation, that the relative distances do not all lie in a plane and that we have to add a third spatial dimension to our model so that two angles are required to specify the direction from any one observer to another. A three-dimensional coordinate system (x, y, z) can be introduced. Adding more observers, we find that no additional spatial dimensions are needed to form a mathematically consistent picture of observations.[1]

VELOCITY AND ACCELERATION

In our space-time model of reality, distance and time are relational quantities. You measure the positions $r(t)$ of a body at various times in relation to some other arbitrary body. You measure the times *t* at which a body is at these positions in relation to the time it was at a specified but arbitrary position. From such measurements you can calculate the time rate of change of position, namely, the velocity,

$$\mathbf{v} = \frac{d\mathbf{r}}{dt} \tag{A5}$$

The above definition is usually taught in class as a more accurate representation of

$$\mathbf{v} = \frac{\Delta \mathbf{r}}{\Delta t} = \frac{\mathbf{r}_2 - \mathbf{r}_1}{t_2 - t_1} \qquad (A6)$$

where $\Delta t \to 0$. Actually, given the ultimately discrete nature of our operational definitions of space and time, infinitesimal calculus is the *approximation*—at least in our model where they are not assumed to have any deeper role.

Similarly, the usual definition of acceleration

$$\mathbf{a} = \frac{d\mathbf{v}}{dt} \qquad (A7)$$

is an approximation to

$$\mathbf{a} = \frac{\Delta \mathbf{v}}{\Delta t} = \frac{\mathbf{v}_2 - \mathbf{v}_1}{t_2 - t_1} \qquad (A8)$$

MASS AND ENERGY

Place a coiled spring between two bodies and measure how much they accelerate after the spring is released (see fig. 2.6, p. 47). Then the ratio of the masses is operationally defined as

$$\frac{m_1}{m_2} = \frac{a_2}{a_1} \qquad (A9)$$

If one of the masses, say, m_2, is taken as a standard, say, one kilogram, then we can measure other masses in relation to this standard.

If a body of constant mass m is observed to have an acceleration $\mathbf{a}$, then the net force on that body is defined as

$$\mathbf{F} = m\mathbf{a} \qquad (A10)$$

which is the familiar form of Newton's second law of motion.

Kinetic energy is the quantity of energy associated with a body's motion. The kinetic energy of a body of mass m moving at a speed $v \ll c$ is

$$K = \frac{1}{2}mv^2 \qquad \text{(A11)}$$

It is important to keep in mind that this formula is *incorrect* at "relativistic" speeds, that is, where v is an appreciable fraction of c.

Potential energy is stored energy that can be converted into other forms of energy. An arbitrary constant can always be added to potential energy without changing any results. However, for the gravitational attraction between bodies, the convention is to assign them zero potential energy when they are separated by infinite distance. Two spherical bodies of masses m_1 and m_2, whose centers are separated by a distance r greater than the sum of their respective radii, have a potential energy

$$V(r) = -\frac{Gm_1m_2}{r} \qquad \text{(A12)}$$

where G is Newton's gravitational constant.

Total mechanical energy is the sum of the kinetic and potential energy of a body

$$E = K + V \qquad \text{(A13)}$$

The rest energy E_0 of a body of mass m is given by the famous formula

$$E_o = mc^2 \qquad \text{(A14)}$$

MOMENTUM AND FORCE

The momentum of a body moving at a speed $v \ll c$ is the product of its mass and velocity,

$$\mathbf{p} = m\mathbf{v} \qquad \text{(A15)}$$

Again, it is important to remember that the above formula must be modified for speeds near the speed of light.

The general form of Newton's second law, that is, the precise definition of force, is

$$\mathbf{F} = \frac{d\mathbf{p}}{dt} \tag{A16}$$

which gives $\mathbf{F} = m\mathbf{a}$ when m is constant and $v \ll c$.

4-MOMENTUM

The mass, energy, and momentum of a body can be collected in a 4-dimensional momentum vector, or 4-momentum $p = (p_0, p_1, p_2, p_3)$ where $p_1 = p_x, p_2 = p_y, p_3 = p_z$, are components of the 3-momentum $\mathbf{p} = (p_x, p_y, p_z)$ and the energy E is included as the zeroth component, p_0, in a way that will be discussed shortly. Note that while boldface is used to designate 3-vectors (familiar 3-space vectors), 4-vectors are represented in normal typeface.

The 4-momentum is not simply a set of four numbers, but a mathematical object that transforms from one coordinate system to another in a way analogous to familiar 3-vectors. The components of two 3-vectors $\mathbf{p}$ and $\mathbf{q}$ usually form orthogonal coordinates in 3-dimensional space, so that the scalar product is invariant to a translation or rotation of the coordinate system in 3-space:

$$\mathbf{p} \cdot \mathbf{q} = pq\cos\theta = \mathbf{p'} \cdot \mathbf{q'} \tag{A17}$$

where θ is the angle between p and q (see fig. A2.2). We can write this

$$p^k q_k = p'^k q'_k \tag{A18}$$

where repeated Latin indices are understood as summed from 1 to 3. Similarly, if the components of two 4-vectors p and q form orthogonal coordinates in 4-dimensional space, the scalar product

$$p^\mu q_\mu = p'^\mu q'_\mu \tag{A19}$$

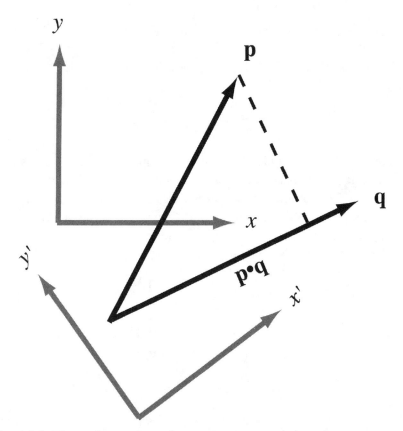

Fig. A2.2. The scalar product of two vectors, **p•q**, is invariant to a translation or rotation of the coordinate system.

where repeated Greek indices are understood as summed from 0 to 3, is invariant to a rotation of the coordinate system in 4-space.

More generally, we write

$$p^\mu g_\mu^\nu q_\nu = p'^\mu g_\mu^\nu q'_\nu \tag{A20}$$

where g_μ^ν is the metric tensor of the space. In Euclidean space, $g_\mu^\nu = \delta_\mu^\nu$, the Kronecker which is 1 for equal indices and 0 for unequal indices.

The metric tensor specifies the geometry, or *metric*, of 4-space and is defined in various ways in physics literature. In higher-level applications I will use the definition that is standard in those applications. For most discussions, I will assume a Euclidean space and take $p_o = \dfrac{iE}{c}$. This

particular definition makes p_o an imaginary number ($i = \sqrt{-1}$). The mass m is given by

$$p^{\mu} g^{\nu}_{\mu} p_{\nu} = -\frac{E^2}{c^2} + \left|\mathbf{p}\right|^2 = -m^2 c^2 \tag{A21}$$

where |p| is the magnitude of the 3-momentum. The mass m is invariant to a rotation in 4-space, which we will show is equivalent to Lorentz invariance. That is, the mass of a particle is the same in all reference frames moving at constant velocity with respect to one another, while the energy and components of 3-momentum are not.

Another frequently used metric takes

$$g = \begin{pmatrix} 1 & 0 & 0 & 0 \\ 0 & -1 & 0 & 0 \\ 0 & 0 & -1 & 0 \\ 0 & 0 & 0 & -1 \end{pmatrix} \tag{A22}$$

In this case, in units $c = 1$,

$$p^{\mu} g^{\nu}_{\mu} p_{\nu} = E^2 - \left|\mathbf{p}\right|^2 = m^2 \tag{A23}$$

which nicely defines the 4-momentum as a 4-vector whose zeroth component is the energy, whose remaining components are the components of the 3-momentum, and whose magnitude is equal to the invariant mass.

SPACE-TIME

The previous discussion on 4-momentum applies to any 4-vector. In the case of space-time (I have reversed the order of the usual discussion), the 4-dimensional position vector is $x = (x_0, x_1, x_2, x_3)$ where $x_1 = x$, $x_2 = y$, and $x_3 = z$ are components of the 3-vector position $\mathbf{r} = (x_1, x_2, x_3)$, and the time t is included as the zeroth component, x_0, in a way that will be discussed below.

In general we can write for two 4-vector positions x and y,

$$x^{\mu} g_{\mu}^{\nu} y_{\nu} = x'^{\mu} g_{\mu}^{\nu} y'_{\nu} \qquad (A24)$$

where g_{μ}^{ν} is the metric tensor described in the previous section. If the Euclidean metric is used, then $x_0 = ict$. In general, the *proper time* τ is given by

$$x^{\mu} g_{\mu}^{\nu} x_{\nu} = -c^2 t^2 + |\mathbf{r}|^2 = -c^2 \tau^2 \qquad (A25)$$

In physics we often run into invariants of this type, the 4-dimensional scalar product of two 4-vectors. For example,

$$p^{\mu} g_{\mu}^{\nu} x_{\nu} = -p^o x_o + \mathbf{p} \bullet \mathbf{r} = -Et + \mathbf{p} \bullet \mathbf{r} \qquad (A26)$$

KICKING BACK IN 4-MOMENTUM SPACE

The basic processes involving particle collisions correspond generically to what I have taken as the archetype of human interaction with reality, kicking it and measuring the return kicks. The basic experiment performed at particle accelerators involves taking a beam of particles of a known mass and carefully calibrated energy and 3-momentum, striking it against a target, and measuring the momenta and energies of the outgoing particles. In the earliest experiments the target comprised a fixed block of matter. In more recent years two beams have been brought into collision, greatly increasing the effective energy of the interaction.

Suppose we have a device that measures the color of each photon that we send out and of each photon that is returned. Our eyes and brains, of course, form such devices. While we usually describe a color measurement in terms of the wavelength of an electromagnetic wave, designated by λ, we have known since the early twentieth century that a focused electromagnetic wave is a beam of photons with a magnitude of 3-momentum $|\mathbf{p}|$ given by

$$|\mathbf{p}| = \frac{h}{\lambda} \qquad (A27)$$

where h is Planck's constant, and energy E where

$$E = |\mathbf{p}|c = \frac{hc}{\lambda} \tag{A28}$$

where c is the speed of light. Thus the 4-momentum of the photon is

$$p = (p_o, p_1, p_2, p_3) = \left(i\frac{E}{c}, p_x, p_y, p_z \right) \tag{A29}$$

Note that the square of the magnitude of the 4-momentum (in the Euclidean metric) is given by

$$-m^2 c^2 = p^\mu p_\mu = -\frac{E^2}{c^2} + |\mathbf{p}|^2 \tag{A30}$$

where m is the mass. For the photon, $m = 0$.

Our basic experiment in 4-momentum space is illustrated in fig. 2.8 (p. 52). Let p_a be the 4-momentum of the particle that corresponds to our kick, and p_b be the 4-momentum of the particle that corresponds to the kickback. The two particles need not, in general, be the same type. Also let us work in units where $c = 1$. The 4-momentum transferred to the struck object is, from 4-momentum conservation,

$$q = p_a - p_b \tag{A31}$$

Squaring this we get, in the Euclidean metric,

$$\begin{aligned} q^2 = \left(p_a - p_b \right)^2 &= -m_a^2 - m_b^2 - 2p_a \bullet p_b \\ &= -m_a^2 - m_b^2 + 2E_a E_b - 2\mathbf{p}_a \bullet \mathbf{p}_b \\ &= -m_a^2 - m_b^2 + 2E_a E_b - 2|\mathbf{p}_a||\mathbf{p}_b|\cos\theta_{ab} \end{aligned} \tag{A32}$$

where θ_{ab} is the angle between the outgoing 3-vector momentum $\mathbf{p}_a$ and the incoming 3-vector momentum $\mathbf{p}_b$. Thus, by measuring the momenta and energies of outgoing and incoming particles, and the angle between them, we can obtain the 4-momentum transfer q^2. Note that I have introduced the spatial concept of angle, but it is the angle in 3-momentum space, which can be defined purely mathematically in terms of momentum components and makes no reference to angles in space-time. That is,

$$\cos\theta_{ab} = \frac{1}{|\mathbf{p}_a||\mathbf{p}_b|}\left(p_{ax}p_{bx} + p_{ay}p_{by} + p_{az}p_{bz}\right)^{1/2} \tag{A33}$$

In general, one has to worry about the spins and other properties of the particles, which are important and add additional information about the object being probed. However, these introduce great complications that require mathematical methods beyond the scope of this book.

Fortunately, the basic idea of working in 4-momentum space can be developed without these complications. In this case, by making many measurements for a fixed p_a we can obtain the probability distribution for a particular momentum transfer. Let $F(q)dq$ be the fraction of measurements that give a momentum transfer q in the range $q \rightarrow q + dq$. Note that $F(q)$ also depends on the particle masses m_a and m_b and the 4-momentum p_a, which are constants. So, the only variable is q.

The job of a model then is to find an expression for $F(q)$ to compare with experiment. This will involve some hypothesis about the nature of the object being struck. That hypothesis will be falsified by the failure to agree with experiments and (tentatively) verified by a successful agreement.

Consider the case where a and b are massless. Then, $E_a = |\mathbf{p}_a|$ and $E_b = |\mathbf{p}_b|$, so

$$q^2 = 2E_aE_b\left(1-\cos\theta_{ab}\right) \tag{A34}$$

is positive. Since $-q^2$ is an invariant mass squared, that mass must be imaginary and cannot correspond to a physically observable particle.

In the interaction scheme described by Feynman diagrams, q is the 4-momentum of a virtual particle X that is emitted by particle a, that then is absorbed by the target particle c in a reaction

$$a + c \rightarrow b + d \tag{A35}$$

as shown in fig. 2.9 (p. 53). In the spinless case, the probability amplitude is simply

$$A(q) = \frac{g_1 g_2}{q^2 + m_X^2} \qquad (A36)$$

where g_1 and g_2 are measures of the strength of the interactions at the two "vertices" in the diagram, and m_X is the mass of the exchanged particle when it is observed as a physical particle. In general, more than one particle can be exchanged and many other topologies (in 4-momentum space) can occur—for example, loops. In principle, we must sum the amplitudes of all Feynman diagrams. The probability distribution will then be the square of the absolute value of that sum.

One specific example of this is Rutherford scattering, where a and b are α-particles, c is the struck nucleus, and d is the recoil nucleus. The α-particle is the nucleus of the helium atom, which is spinless and has an electric charge $2e$. In the case, X is the photon, so $m_X = 0$, $g_1 = 2e$, $g_2 = Ze$, where Z is the atomic number of the struck nucleus. Thus,

$$A(q) = \frac{2Z\,e^2}{q^2} \qquad (A37)$$

In the approximation where this diagram dominates,

$$F(q) = \frac{4Z^2 e^4}{q^4} \qquad (A38)$$

which agrees with experiment.

Of course, Rutherford scattering was the process by which Lord Rutherford inferred the nuclear model of the atom. That model is expressed in space-time language and affords us the opportunity to discuss the connection between the space-time and energy-momentum pictures. Consider the Fourier transform of the amplitude $A(q)$:

$$\overline{A}(x) = \int A(q)\exp(-iq \bullet x)d^4q \qquad (A39)$$

Calculating this for the generic example above, where the X particle can have nonzero mass,

$$\overline{A}(x) = g_1 g_2 \int \frac{1}{q^2 + m_X^2}\exp(-iq \bullet x)d^4q \qquad (A40)$$

Thus, the two pictures are complementary with space-time variables in one case and energy-momentum variables in the other. This is, of course, a manifestation of the *complementarity* of quantum mechanics that is often confusingly referred to as the *wave particle duality*. Note that in neither case is the wave picture used; in both cases we have particles interacting with particles.

NOTE

1. Although additional spatial dimensions are assumed in string theory.

MATHEMATICAL SUPPLEMENT B

CLASSICAL MECHANICS AND RELATIVITY

GALILEO'S EXPERIMENT

Let us look at Galileo's experiment in which he dropped cannonballs and other objects from various heights, according to legend, from the Leaning Tower of Pisa. Suppose we let y be the vertical coordinate and take the coordinate origin $y = 0$ to be at Galileo's position at some balcony in the tower. Let us start our clock when Galileo releases the object, that is, $y = 0$ when $t = 0$. Making measurements of the object's position at various times, we determine that they are fit by the equation

$$y(t) = -\frac{1}{2}gt^2 \qquad \text{(B1)}$$

where g = 9.8 meters per second per second, independent (to a first approximation) of the mass and other properties of the object.

Now, this equation describes Galileo's observations. But is it a universal law of physics? No, for many reasons. Mentioning just one, we can change the origin of the y-coordinate system to be, say, on the ground. Then, when the object is dropped from a height h,

$$y(t) = h - \frac{1}{2}gt^2 \qquad \text{(B2)}$$

Further, suppose we start our clock five seconds before we drop the object. Then, for $t \geq 5$,

$$y(t) = h - \frac{1}{2}g(t-5)^2 \tag{B3}$$

So we have a different equation for each situation. They are not wrong, just not suitable for universal laws. However, if we write

$$y(t) = y(t_o) - \frac{1}{2}g(t-t_o)^2 \tag{B4}$$

we have an equation that is good for $t \geq t_o$ and whatever time t_o we start our clock and from whatever position $y(t_o)$ we drop the object. But this is still not sufficient for universality.

If the laws of physics cannot depend on such a trivial matter as where we place the origin of our coordinate system, then they surely cannot depend on how we orient the axes of that system. For example, suppose we rotate the previous system (x, y) counterclockwise by an angle θ. Then we would have

$$x(t) = x(t_o) - \frac{1}{2}g\cos\theta(t-t_o)^2$$
$$y(t) = y(t_o) - \frac{1}{2}g\sin\theta(t-t_o)^2 \tag{B5}$$

In elementary physics, the equations describing the motion of projectiles are made independent of the choice of coordinate system by using *vectors*. We write the vector equation

$$\mathbf{r}(t) = \mathbf{r}(t_o) + \mathbf{v}(t_o)(t-t_o) - \frac{1}{2}\mathbf{g}(t-t_o)^2 \tag{B6}$$

where r(t) is the vector position of the body at time t, $\mathbf{v}(t_o)$ is the vector velocity at time t_o, and **g** is the vector acceleration due to gravity, pointing to the center of Earth. We can tentatively call this the *universal law of projectiles*. However, as we will see below, simply satisfying the cosmological principle is not a sufficient criterion for universality.

But first let us generalize these ideas. Suppose we specify the position of a body in space at a given time t by a vector $\mathbf{r} = (x, y, z)$. We then have some equation, some "law," that describes the motion of that body as a function of position and time, t,

$$f(x, y, z, t) = K \tag{B7}$$

where K is a constant. The cosmological principle requires that this function be the same for any other coordinate system displaced in position or time

$$f(x', y', z', t') = K \tag{B8}$$

Suppose that the spatial coordinates are shifted by fixed amounts:

$$x' = x + a \quad y' = y + b \quad z' = x + c \tag{B9}$$

If the function is unchanged, we say that it possesses *space-translation invariance.*

If the time coordinate is similarly shifted by a constant,

$$t' = t + d \tag{B10}$$

and the function is unchanged, we say that it possesses *time-translation invariance.* The cosmological principle is thus implemented by guaranteeing that the equations describing a law of physics possess space-translation and time-translation invariance.

Let us rotate our coordinate system by an angle θ around the z-axis,

$$\begin{aligned} x' &= x\cos\theta - y\sin\theta \\ y' &= x\sin\theta + y\cos\theta \end{aligned} \tag{B11}$$

If the function is unchanged by such a rotation, it possesses *space-rotation invariance.* Rotations about other axes can be handled in a similar fashion. While not normally mentioned as part of the cosmological principle, space-rotation invariance clearly belongs there.

Thus, Newton's second law of motion can be written

$$\mathbf{F} - m\frac{d^2\mathbf{r}}{dt^2} = 0 \tag{B12}$$

Since a vector has the same magnitude and direction no matter how you move the coordinate system around,

$$\mathbf{F'} - m\frac{d^2\mathbf{r'}}{dt^2} = 0 \qquad \text{(B13)}$$

where $dt' = dt$. Thus, Newton's law is invariant under space translations, time translations, and space rotations.

GALILEO'S PRINCIPLE OF RELATIVITY

Galileo's principle of relativity states that the laws of physics cannot depend on an absolute velocity. Let us apply this principle to the Galileo experiment. Suppose an observer is riding on a boat that is moving in a straight line at a constant speed v on a river that runs through Pisa near where Galileo is making his observations. Further suppose we define a coordinate system (x', y') attached to the ground where x' is horizontal in the direction of the motion of the boat and y' is vertical, up, with the origin $(0,0)$ at the observer in the boat. Let $t = 0$ be the time at which the object is released from a height h. Then measurements of the falling object's position as a function of time will give

$$x'(t) = -vt$$
$$y'(t) = h - \frac{1}{2}gt^2 \qquad \text{(B14)}$$

To obey Galileo's principle of relativity, the same function must apply in any coordinate system (x', y') that is moving at a constant velocity with respect to (x, y). Suppose that the new system is moving at a speed v along the x-axis of the original system. (Assuming space-translation, space-rotation, and time-translation invariance we can choose any coordinate axes we want). Then,

$$x' = x - vt = -vt$$
$$y' = y = h - \frac{1}{2}gt^2 \qquad \text{(B15)}$$

Eliminating t, we get

$$y' = h - \frac{1}{2}g\left(\frac{x'}{v}\right)^2 \tag{B16}$$

which is the equation for a parabola. That is, the observer on the boat sees the object fall along a parabolic path (see fig. 3.1, p. 60).

In general, we can write

$$x' = x - vt \quad y' = y \quad z' = z \tag{B17}$$

This is called the *Galilean transformation*. These equations take you from (x, y, z) to (x', y', z'), or vice versa. The principle of relativity then requires that all the equations of physics be invariant to a Galilean transformation.

Note that time is assumed to be invariant, that is, $t' = t$. Einstein showed that the Galilean transformation must be modified to allow for different times in two reference frames. Let us assume for now that $v \ll c$, where this effect is small.

Galilean invariance requires that

$$f(x', y', z', t') = f(x, y, z, t) \tag{B18}$$

For example, Newton's second law of motion can be written

$$F_x - m\frac{d^2x}{dt^2} = F'_x - m\frac{d^2x'}{dt'^2} = 0 \tag{B19}$$

with similar equations for F_y and F_z. Clearly,

$$\frac{d^2x'}{dt'^2} = \frac{d^2x}{dt^2} \tag{B20}$$

with similar expressions for y' and z', so the second law is Galilean invariant.

Contrast this with what might be called "Aristotle's law of motion," where a force is necessary to produce a velocity:

$$F_x - k\frac{dx}{dt} = 0 \tag{B21}$$

where k is a constant. Note that

$$\frac{dx'}{dt'} = \frac{dx}{dt} - v \tag{B22}$$

so Aristotle's law in the primed coordinate system reads, assuming $F_{x'} = F_x$ (for example, the force could be zero),

$$F_{x'} - k\frac{dx'}{dt'} - kv = 0 \tag{B23}$$

and Aristotle's law is not invariant to a Galilean transformation.

What is the simplest equation for the vector force **F** that is proportional to the mass of the body (assumed constant) and derivatives of the body's position vector **r** with respect to time? The equation must be invariant under space translation, space rotation, time translation, and a Galilean transformation. Assume $v \ll c$. The answer would have to be:

$$\mathbf{F} = m\frac{d^2\mathbf{r}}{dt^2} \tag{B24}$$

As we have seen, the first derivative would not be Galilean invariant. Higher derivatives would be invariant, as would any linear combination of derivatives of orders greater than the first. But the above form is the simplest consistent with the assumed invariance principles.

FICTITIOUS FORCES

Consider two coordinate systems accelerating with respect to one another. For example, let a be the acceleration of coordinate x' with respect to coordinate x.

$$\frac{d^2x'}{dt'^2} = \frac{d^2x}{dt^2} + a \tag{B25}$$

If

$$F_x = m\frac{d^2x}{dt^2} \tag{B26}$$

then, assuming the force is the same in both reference frames (again, we can simply take the force to be zero),

$$F_{x'} = m\frac{d^2x'}{dt'^2} - ma \tag{B27}$$

That is, Newton's second law is apparently violated in the primed coordinate system.

We can restore the second law by adding the fictitious force ma to the left-hand side of the equation. That is, we write

$$F_{x'}^{\mathit{eff}} = F_{x'} + ma = m\frac{d^2x'}{dt'^2} \tag{B28}$$

where $F_{x'}^{\mathit{eff}}$ is the effective force, namely, the force that equals the mass times the acceleration of the body in the given reference frame.

LAGRANGIAN MECHANICS

Let us define generalized spatial coordinates $q = \{q_1, q_2, \ldots, q_n\}$, which could be the familiar Cartesian coordinates or other types of spatial coordinates, such as angles of rotation about various axes. The *Lagrangian* is defined (in most familiar cases) as the difference in the total kinetic and potential energies of the system

$$L(q_k, \frac{dq_k}{dt}) = T(q_k, \frac{dq_k}{dt}) - V(q_k) \tag{B29}$$

where $k = 1, \ldots, n$; $\frac{dq_k}{dt}$ is the velocity associated with the degree of freedom k; and V is assumed to be independent of $\frac{dq_k}{dt}$. (For the more general situation, see textbooks in graduate-level classical mechanics such as Goldstein.)[1] The equations of motion are then determined by Lagrange's equations

$$\frac{d}{dt}\left(\frac{\partial L}{\partial \frac{dq_k}{dt}}\right) - \frac{\partial L}{\partial q_k} = 0 \tag{B30}$$

Associated with each q_k is a *canonical momentum*, defined as

$$p_k = \frac{\partial L}{\partial \frac{dq_k}{dt}} \tag{B31}$$

In the common special case where T depends only on $\frac{dq_k}{dt}$.

$$\frac{dp_k}{dt} = -\frac{\partial V}{\partial q_k} = F_k \tag{B32}$$

which is Newton's second law.

HAMILTONIAN MECHANICS

Let us define a *Hamiltonian H* that, in most applications, is simply the total energy of the system,

$$H = T(q_k, p_k) + V(q_k, p_k) \tag{B33}$$

where $p = (p_1, p_2, ..., p_n)$. Hamilton's equations of motion then are

$$\frac{dp_k}{dt} = -\frac{\partial H}{\partial q_k} \tag{B34}$$

where $k = 1, \ldots, n$, which again can be seen to give Newton's second law of motion when $V = V(q_k)$, that is, is independent of p_k, and

$$\frac{dq_k}{dt} = \frac{\partial H}{\partial p_k} \tag{B35}$$

which gives the velocity corresponding to coordinate q_k.

LEAST ACTION

In yet another generalized method, we define a quantity called the *action*,

$$S = \int_{A \to B} L dt \tag{B36}$$

where the integral is over some path from a point A to a point B. It can be shown that the path predicted by the equations of motion is precisely the path for which the action is *extremal*, that is, either maximum or minimum.

THE CLOCKWORK UNIVERSE

Consider the most elementary application of Newton's second law of motion, $\mathbf{F} = m\mathbf{a}$, where we have some force $\mathbf{F}$ acting on a body of constant mass m. If we know how $\mathbf{F}$ depends on time, we can integrate once to get the velocity as a function of time,

$$\mathbf{v}(t) = \frac{1}{m}\int_0^t \mathbf{F}(t)dt + \mathbf{v}(0) \qquad (B37)$$

and a second time to get the position as a function of time,

$$\mathbf{r}(t) = \int_0^t \mathbf{v}(t)dt + \mathbf{r}(0) \qquad (B38)$$

Thus we are able to predict the future motion of the particle from a knowledge of the initial position, $\mathbf{r}(0)$, the initial velocity, $\mathbf{v}(0)$, and the force as a function of time, $\mathbf{F}(t)$.

MAXWELL'S EQUATIONS

The following are Maxwell's equations in the Standard International system of units, for the electric displacement vector $\mathbf{E}$ and the magnetic induction $\mathbf{B}$ in a vacuum, where ρ is the electric charge density, $\mathbf{J}$ is the electric current density, ε_o is the electric permittivity of free space, and μ_o is the magnetic permeability of free space. Consult any advanced text on electrodynamics for the more general case of fields inside normal material media.

Gauss's law of electricity

$$\nabla \bullet \mathbf{E} = \frac{\rho}{\varepsilon_o} \tag{B39}$$

Faraday's law of induction

$$\nabla \times \mathbf{E} = -\frac{\partial \mathbf{B}}{\partial t} \tag{B40}$$

Gauss's law of magnetism

$$\nabla \bullet \mathbf{B} = 0 \tag{B41}$$

Ampère's law with Maxwell's modification

$$\nabla \times \mathbf{B} = \mu_o \mathbf{J} + \mu_o \varepsilon_o \frac{\partial \mathbf{E}}{\partial t} \tag{B42}$$

Once **E** and **B** are calculated, the *Lorentz force equation*

$$\mathbf{F} = q(\mathbf{E} + \mathbf{v} \times \mathbf{B}) \tag{B43}$$

is used to compute the force on a particle of charge q. And once we have the force, we can use the equations of motion to predict the motion of the particle under the action of electric and magnetic fields.

   Maxwell had added the *displacement current* term $\mathbf{J_D} = \varepsilon_o \dfrac{\partial \mathbf{E}}{\partial t}$

to Ampère's law. This made it possible for the fields to sustain one another, as long as they varied with time. Solving Maxwell's equations at a position in empty space where no charges or currents exist, that is, $\rho = 0$ and $\mathbf{J} = 0$, one gets the wave equations

$$\nabla^2 \mathbf{E} - \mu_o \varepsilon_o \frac{\partial^2 \mathbf{E}}{\partial t^2} = 0 \tag{B44}$$

and

$$\nabla^2 \mathbf{B} - \mu_o \varepsilon_o \frac{\partial^2 \mathbf{B}}{\partial t^2} = 0 \tag{B45}$$

That is, $\mathbf{E}$ and $\mathbf{B}$ each propagate as a wave with a speed $c = (\mu_o \varepsilon_o)^{-1/2} = 3 \times 10^8$ meters per second. This is exactly the speed of light in a vacuum.

THE LORENTZ TRANSFORMATION

Maxwell's equations are not Galilean invariant. The easiest way to see this is to look at the wave front of a spherical electromagnetic wave emitted from a point. Spherical waves comprise one solution of Maxwell's equations. The wave front will be given by the equation

$$c^2 t^2 - x^2 - y^2 - z^2 = 0 \qquad (B46)$$

By doing a Galilean transformation while demanding the speed of light be unchanged we get

$$c^2 t^2 - (x' + vt)^2 - y'^2 - z'^2 = 0 \qquad (B47)$$

which is clearly not invariant, depending on v.

Lorentz discovered that Maxwell's equations were invariant under the transformation

$$x' = \gamma(x - vt)$$
$$y' = y$$
$$z' = z \qquad (B48)$$
$$t' = \gamma\left(t - \frac{vx}{c^2}\right)$$

where $\gamma = \dfrac{1}{\sqrt{1 - \dfrac{v^2}{c^2}}}$ called the Lorentz factor. Note that time also enters

with a transformation equation of its own.

It is important to also note the inverse transformation, for which $v \rightarrow -v$:

$$x' = \gamma(x - vt)$$
$$y' = y$$
$$z' = z \qquad\qquad\qquad \text{(B49)}$$
$$t' = \gamma\left(t - \frac{vx}{c^2}\right)$$

A little algebra shows that

$$c^2t^2 - x^2 - y^2 - z^2 = c^2t'^2 - x'^2 - y'^2 - z'^2 \qquad \text{(B50)}$$

proving that the spherical wave front is Lorentz invariant.

It follows from the Lorentz transformation that moving clocks appear to slow down (*time dilation*) and that a moving body appears to contract along its direction of motion (*Lorentz-Fitzgerald contraction*). These are special cases of the more general fact implied by the Lorentz transformation: *The distance intervals measured with a meterstick and the time intervals measured with a clock will be, in general, different for two observers moving relative to one another.*

Hermann Minkowski introduced the notion of 4-dimensional space-time, where time is one of the four coordinates. A number of different conventions are still in use. In the simplest, the position of a point in space-time is given by the 4-vector $x = (x_o, x_1, x_2, x_3) = (ict, x, y, z)$ where $i = \sqrt{-1}$. Recall from chapter 1 that the 4-vector momentum in this convention is

$$p = (p_o, p_1, p_2, p_3) = (i\frac{E}{c}, p_x, p_y, p_z) \qquad \text{(B51)}$$

where E is the energy and $\mathbf{p} = (p_x, p_y, p_z)$ is the familiar 3-momentum. In relativistic physics, casting variables in 4-vector form makes Lorentz invariance clear.

Maxwell's equations are invariant in form to a Lorentz transformation, although the fields $\mathbf{E}$ and $\mathbf{B}$ are not themselves invariant. These fields cannot be expressed as 4-vectors. However, the $\mathbf{E}$ and $\mathbf{B}$ fields can be determined from a scalar potential ϕ and a vector potential $\mathbf{A}$ that can be made part of a 4-potential $A = (i \dfrac{\phi}{c}, A_x, A_y, A_z)$. Similarly, the electric charge and the current densities can be written in terms of a 4-current density $J = (i\rho c \, J_x, J_y, J_z)$, where ρ is the charge density.

Finally, Maxwell's equation can be expressed in elegant tensor form, which directly manifests their Lorentz invariance:

$$\frac{\partial F_{\mu\nu}}{\partial x_\nu} = \mu_o J_\mu \tag{B52}$$

and

$$\frac{\partial G_{\mu\nu}}{\partial x_\nu} = 0 \tag{B53}$$

where

$$F_{\mu\nu} = \begin{pmatrix} 0 & E_x & E_y & E_z \\ -E_x & 0 & B_z & -B_y \\ -E_y & -B_z & 0 & B_x \\ -E_z & B_y & -B_x & 0 \end{pmatrix} \tag{B54}$$

and

$$G_{\mu\nu} = \begin{pmatrix} 0 & B_x & B_y & B_z \\ -B_x & 0 & -E_z & E_y \\ -B_y & E_z & 0 & -E_x \\ -B_z & -E_y & -E_x & 0 \end{pmatrix} \tag{B55}$$

where $c = 1$ and repeated Greek indices are summed from 0 to 4.

GENERAL RELATIVITY

Now, let us see how the principle of covariance together with the Leibniz-Mach and equivalence principles leads to general relativity. Consider the equation of motion for a freely falling body in terms of a coordinate system, $y = (y_0, y_1, y_2, y_3)$, where $y_0 = ict$, falling along with the body.

Since the body does not change position in its own reference frame, the 3-vector $d\mathbf{y}/dt = 0$. It also does not accelerate, so

$$\frac{d^2\mathbf{y}}{dt^2} = 0 \tag{B56}$$

Also, since, $y_o = ict$,

$$\frac{d^2 y_o}{dt^2} = 0 \tag{B57}$$

Furthermore,

$$(d\tau)^2 = (dt)^2 - \frac{1}{c^2}\left|d\mathbf{y}\right|^2 = (dt)^2 \tag{B58}$$

Let us work in units where $c = 1$. We can write the above in 4-vector form,

$$\frac{d^2 y_\alpha}{d\tau^2} = 0 \tag{B59}$$

where $\alpha = 0, 1, 2, 3$. This is the 4-dimensional equation of motion for a body acted on by zero force. This expresses the fact that a freely falling body experiences no external force.

Next, let us consider a second coordinate system x_α fixed to a second body such as Earth. This could be any coordinate system accelerating with respect to the first. The equation of motion can be transformed to that coordinate system as follows:

$$\frac{d}{d\tau}\left(\frac{\partial y_\rho}{\partial x_\mu}\frac{dx_\mu}{d\tau}\right) = 0 \tag{B60}$$

from which, after some algebra,[2] we find

$$\frac{d^2 x_\lambda}{d\tau^2} + \Gamma_{\lambda\mu\nu}\frac{dx_\mu}{d\tau}\frac{dx_\nu}{d\tau} = 0 \tag{B61}$$

where

$$\Gamma_{\lambda\mu\nu} = \frac{\partial x_\lambda}{\partial y_\rho}\frac{\partial^2 y_\rho}{\partial x_\mu \partial x_\nu} \tag{B62}$$

is called the *affine connection*. An observer on Earth witnesses a body accelerating toward Earth and interprets it as the action of a "gravitational force." Although $\Gamma_{\lambda\mu\nu}$ has three indices, it is not a tensor since it is not Lorentz invariant.

We can obtain the Newtonian picture in the limit of low speeds, $dx_k/dt \approx 0$. In this case, $d\tau = dt$, $d^2x_o/dt^2 = 0$, and

$$\frac{d^2x_k}{dt^2} = -\Gamma_{koo} = g_k \tag{B63}$$

for $k = 1, 2, 3$, where $\mathbf{g} = (g_1, g_2, g_3)$ is the Newtonian field vector. Thus, the Γ_{koo} elements of the affine connection are just the negatives of the Newtonian gravitational field components in the limit of low speeds. Additional elements then are needed to describe gravity at speeds near the speed of light.

The Newtonian field vector for any distribution of mass can be obtained from the gravitational potential ϕ, which is in general a solution of Poisson's equation

$$\nabla^2\phi = 4\pi G\rho \tag{B64}$$

where ρ is the mass density and $\mathbf{g} = -\nabla\phi$. For example, suppose that we have a point mass m so that $\rho(\mathbf{r}) = m\delta(0)$. Then,

$$\nabla^2\phi = 4\pi G\rho \tag{B65}$$

and

$$\mathbf{g} = -\nabla\phi = -\frac{Gm}{r^2}\hat{\mathbf{r}} \tag{B66}$$

is the familiar Newtonian result.

While the modified equation of motion above contains relativistic effects of gravity, it is not Lorentz invariant. It has a different form in the two reference frames. Einstein sought to find equations to describe gravity that were Lorentz invariant. He started with the Poisson equation above, which is now non-Lorentz invariant since ρ is the mass or energy density, and where energy is the zeroth component of a 4-vector, which is also not Lorentz invariant.

Let us follow Einstein and search for an invariant quantity to replace density. Suppose we have a dust cloud in which all the dust particles are moving slowly, that is, with $v \ll c$, in some reference frame. Let the energy density in that frame be ρ_0. Let E_0 be the rest energy of each particle ($c = 1$) and n_0 be the number per unit volume. Then,

$$\rho_0 = E_0 n_0 \tag{B67}$$

In some other reference frame the energy density will be

$$\rho = En = \gamma E_0 \gamma n_0 = \gamma^2 n_0 \tag{B68}$$

where γ is the Lorentz factor, $E = \gamma E_0$, and $n = \gamma n_0$. To see the latter, note that $n = dN/dV$, where dN is the number of particles in the volume element dV, $dN_0 = dN$, and $dV = dV_0/\gamma$ from the Fitzgerald-Lorentz contraction.

Also note that ρ is not simply the component of a 4-vector, because of the factor γ^2. Rather it must be made part of a second-rank tensor. We can write

$$T_{\mu\nu} = \rho_0 v_\mu v_\nu \tag{B69}$$

where v_μ is the 4-velocity of the cloud. Then, since $v_0 = d\tau/dt = \gamma$,

$$T_{00} = \rho_0 v_0 v_0 = \gamma^2 \rho_0 = \rho \tag{B70}$$

That is, T_{00} is just the density. $T_{\mu\nu}$ is the *energy-momentum tensor*, or the *stress-energy tensor*. The other components comprise energy and momentum flows in various directions: T_{0i} is the energy flow per unit area in the *i*-direction, for example, a heat flow; T_{ii} is the flow of momentum component *i* per unit area in that direction, the pressure across the *i*-plane; T_{ij} is the flow of momentum component *i* per unit area in the *j*-direction, the viscous drag across the *j*-plane; and T_{i0} is the density of the *i*-component of momentum.

Einstein thus wrote, as the invariant form of Poisson's equation,

$$G_{\mu\nu} = -8\pi G T_{\mu\nu} \tag{B71}$$

where G is Newton's constant and the factor is chosen so that we get Poisson's equation in the Newtonian limit. Since the energy-momentum tensor $T_{\mu\nu}$ is invariant, the quantity $G_{\mu\nu}$ is also an invariant tensor field.

In what has become the standard model of general relativity, Einstein related $G_{\mu\nu}$ to the curvature of space-time in a non-Euclidean geometry. This assured that all bodies would follow a geodesic in space-time in the absence of forces, the gravitational force being thus eliminated and replaced by a curved path. In non-Euclidian geometry, the proper distance Δs between two points in space-time is given by

$$(\Delta s)^2 = \Delta x_\mu g_{\mu\nu} \Delta x_\nu \tag{B72}$$

where $g_{\mu\nu}$ is the metric tensor.

Einstein assumed that $G_{\mu\nu}$ is a function of $g_{\mu\nu}$ and its first and second derivatives. In its usual form, *Einstein's field equation* is given as

$$R_{\mu\nu} - \frac{1}{2} g_{\mu\nu} R + \Lambda g_{\mu\nu} = -8\pi G T_{\mu\nu} \tag{B73}$$

where $R_{\mu\nu}$ and R are contractions of the rank four Riemann curvature tensor. To see the explicit forms of these quantities, consult any textbook on general relativity. The quantity Λ is the infamous *cosmological constant*.

NOTES

1. Herbert Goldstein, *Classical Mechanics*, 2nd ed. (New York: Addison-Wesley, 1980).

2. Steven Weinberg, *Gravitation and Cosmology: Principles and Applications of the General Theory of Relativity* (New York: Wiley, 1972), p. 102.

3. Ibid., pp. 151–57.

MATHEMATICAL SUPPLEMENT C

INTERACTION THEORY

RADIATION RESISTANCE

Classical electrodynamics predicts that an accelerated charged particle will emit radiation and consequently lose energy. In order to conserve energy, work must be done on the particle to account for that energy loss. The force that does this work is called the *radiation resistance*, or the *radiation damping force*, and opposes any external forces that are acting to accelerate the particle. However, if the particle is all by itself, with no external forces present other than the force producing the acceleration, then what is doing this resistive work? The standard answer was that the particle does work on itself. But how?

Radiation resistance was interpreted by Lorentz and other nineteenth-century physicists as a *self-action* in which the radiation emitted by the particle acts back on the particle itself. Electromagnetic radiation carries momentum, so if radiation is being emitted forward by a particle, that particle must necessarily recoil backward to conserve momentum.

Consider two particles, A and B, of equal mass and charge forming a rigid body like a dumbbell. The total mass is m_o and the total charge is q. Suppose the body is being accelerated along the line between the particles by some external force F_{ext} such as a static electric field. The motion of the body as a whole can be treated as if it were a single mass acted on by whatever external forces are present. That is,

$$F_{ext} + F_A + F_B = 2\left(\frac{m_o}{2}a\right) = m_o a$$

$$F_A = -F_B \tag{C1}$$

$$F_{ext} = m_o a$$

where the internal forces cancel by Newton's third law.

However, Lorentz recognized that this would not be the case when the finite propagation speed of electromagnetic waves was taken into account. Imagine our two particles within the body initially at rest. Particle A emits a pulse of electromagnetic radiation in the direction of particle B, which we will take to be the direction of the acceleration of the system. Simultaneously, B emits a pulse in the direction of A. Each pulse moves at the speed of light and carries a certain momentum dp that is to be transferred to the struck particle.

Although the particles start at rest, they begin moving as the result of their acceleration. Let dt_A be the time between pulses emitted by A. Then the force on A is opposite to the direction of the emitted pulse,

$$F_A = -\frac{dp}{dt_A} \tag{C2}$$

In the time it takes the pulse to go from A to B, B will have moved relatively farther from the point where A emitted its pulse. Thus, the time between pulses received by B, dt_B, will be greater than dt_A. It follows that

$$F_B = \frac{dp}{dt_B} < -F_A \tag{C3}$$

so

$$F_A + F_B \equiv F_{rr} < 0 \tag{C4}$$

and

$$F_{ext} + F_{rr} = m_o a \tag{C5}$$

where F_{rr} is the radiation resistance that opposes the external force F_{ext}. This resistive force results from the finite speed of light.

Indeed, you can turn the picture around and argue that the accelerated charge would not emit any radiation if it where not for this difference between the internal forces that results from the finite speed of light. Since the internal work done by those forces does not cancel, some energy must be emitted.

The force of radiation resistance depends on the acceleration a. Let us assume it can be expanded in a power series,

$$F_{rr} = -m_s a - ba^2 - da^3 - \ldots \tag{C6}$$

where I have explicitly indicated that the force is negative (although b, d, and any higher-order coefficients can have either sign as long as the net is negative). Then we can write

$$F_{ext} + ba^2 + da^3 + \ldots = \left(m_o + m_s\right)a = ma \tag{C7}$$

The term proportional to acceleration has been taken over to the other side of $F = m_o a$, the $m_o a$ side, resulting in an effective increase in the mass of the body. We call this added mass m_s the *self-mass*.

The self-mass, which reflects the increased inertia that a charged particle exhibits when accelerated, was first calculated by J. J. Thomson in 1881.[1] We can provide a simple derivation using special relativity. The added mass is just the electrical potential energy of the body, the repulsive energy that is ready to blow the particle apart, divided by the speed of light squared. That is,

$$m_s c^2 = \frac{1}{4\pi\varepsilon_o} \frac{\left(q/2\right)^2}{d} \tag{C8}$$

where q is the total charge of the body and d is the separation between the two particles. Unhappily, the electromagnetic self-mass goes to infinity when the diameter of the body goes to zero, that is, when the body becomes a point particle.

SUBTRACTING INFINITY

In 1938 Hendrick Kramers realized that the electromagnetic self-mass was being added to the particle's "intrinsic mass," a quantity that is never actually measured. As we saw above, this added mass results from taking that part of the radiation reaction that is proportional to acceleration and moving it over to the other side of $F = ma$, that is, including it in the mass m. What exactly is the m in $F = ma$? It is not some abstract, metaphysical "bare" mass of a particle. Rather, m is the inertial mass that we measure in an experiment when we apply a measured force F to a particle and then measure the resulting acceleration a. That is, $m = F/a$ is an operational quantity.

Kramers suggested that the self-mass of a charged particle is simply absorbed in its measured inertial mass. Thus, when you do physics, such as when you put a charged particle in an electric field, the self-mass is already included in the measured mass. The fact that someone calculates infinity for the self-mass in his model is an artifact of his model.

Since Kramers made his suggestion within the framework of classical physics, its significance was not immediately recognized. However, he presented his proposal ten years later at the 1948 Shelter Island conference that marked the starting point of the explosive developments in postwar quantum electrodynamics (QED), for which renormalization was the key.[2] The renormalization program that was then instituted for QED followed the lines that Kramers had outlined. The infinities that people kept calculating were recognized as of their own doing. They did not appear in observations, and so they were clearly wrong. They had to be moved over to the other side of the equation and subsumed in what we actually observe in nature.

THE FEYNMAN-WHEELER INTERACTION THEORY

In ordinary classical electrodynamics, the electric field $\mathbf{E}(\mathbf{r},t)$ and the magnetic field $\mathbf{B}(\mathbf{r},t)$ detected at a given position $\mathbf{r}$ and time t are calculated from the distribution of charges $\rho(\mathbf{r}',t_r)$ and currents $\mathbf{J}(\mathbf{r}',t_r)$ at the source at an earlier time called the *retarded time*,

$$t_r = t - \frac{|\mathbf{r} - \mathbf{r}'|}{c} \tag{C9}$$

The potentials from which **E** and **B** can be calculated are

$$V(\mathbf{r},t) = \frac{1}{4\pi\varepsilon_o} \int \frac{\rho(\mathbf{r}',t_r)}{|\mathbf{r}-\mathbf{r}'|} d^3\mathbf{r}' \tag{C10}$$

and

$$\mathbf{A}(\mathbf{r},t) = \frac{1}{4\pi\varepsilon_o} \int \frac{\mathbf{J}(\mathbf{r}',t_r)}{|\mathbf{r}-\mathbf{r}'|} d^3\mathbf{r}' \tag{C11}$$

However, like all the basic laws of physics, Maxwell's equations are time symmetric—they do not single out a particular direction of time. Thus, the same results can be obtained by computing the distribution of charges and currents at the source at a later time, called the *advanced time*,

$$t_a = t + \frac{|\mathbf{r}-\mathbf{r}'|}{c} \tag{C12}$$

where the potentials are

$$V(\mathbf{r},t) = \frac{1}{4\pi\varepsilon_o} \int \frac{\rho(\mathbf{r}',t_a)}{|\mathbf{r}-\mathbf{r}'|} d^3\mathbf{r}' \tag{C13}$$

and

$$\mathbf{A}(\mathbf{r},t) = \frac{1}{4\pi\varepsilon_o} \int \frac{\mathbf{J}(\mathbf{r}',t_a)}{|\mathbf{r}-\mathbf{r}'|} d^3\mathbf{r}' \tag{C14}$$

That is, you can equivalently treat the problem as if the signal arrives at the detector before it leaves the source. This solution is conventionally ignored by making the assumption, not usually recognized as an assumption, that cause must always precede effect. In practical problems, this is usually built into the boundary conditions.

In the early 1940s Feynman and Wheeler hypothesized that charged particles do not act on themselves, only on each other. However, the motion of a charge is not only determined by the *previous* motion of other charges, but by their *later* motion as well. Furthermore, all the particles in the Universe must be considered in understanding the motion of one particle.

Let us go back to the situation described earlier in which the self-force of a two-particle system was seen to result from the differing time delays between the emission and the absorption of the electromagnetic pulses sent from one particle to the other. The different delays in the two directions resulted in a net force opposite to the direction of acceleration that was interpreted as the radiation resistance. Now imagine what happens if we include effects from the *future* as well as the past, as we are logically required to do by the time-symmetry of electrodynamics. Then,

$$F_{rr} = F_{A,ret} + F_{B,ret} + F_{A,adv} + F_{A,adv}$$
$$F_{A,ret} = -F_{A,adv} \tag{C15}$$
$$F_{B,ret} = -F_{B,adv}$$

From which it follows that

$$F_{rr} = 0 \tag{C16}$$

That is, the pulses from the advanced time, the "future," will exactly cancel those from the retarded time, the "past," leaving a zero net self-force. Instead of being infinite, the electromagnetic correction to the electron's inertial mass is zero!

But if the electromagnetic self-mass is zero, we still have to account for radiation. Consider the interaction between two electrons A and B separated by some distance. Suppose electron A is set oscillating with a radial frequency ω at a time T_1. It will emit an electromagnetic wave that, provided the separation distance is many wavelengths $\lambda = 2\pi c/\omega$, will be well approximated by the plane wave

$$E_{A,ret} = e^{i(kx - \omega t)} \tag{C17}$$

for $t \geq T_1$, where we take the amplitude to be unity and $k = \omega/c$. We call this the *retarded field*. It is zero for $t \leq T_1$.

Let us assume, consistent with the basic time symmetry of Maxwell's equations, that the electron also emits an *advanced wave* in the opposite time direction,

$$E_{A,adv} = e^{-i(kx - \omega t)} \tag{C18}$$

for $t \geq T_1$.

The retarded field propagates along the *x*-direction until it encounters electron B at a time $t = T_2$. This electron is then set oscillating with the same frequency, producing the retarded wave

$$E_{B,ret} = -e^{i(kx - \omega t)} \qquad (C19)$$

for $t \geq T_2$, where, if the wave is completely absorbed,

$$E_{A,ret} + E_{B,ret} = 0 \qquad (C20)$$

The retarded wave is zero for $t \geq T_2$.

Similarly, an advanced wave is produced.

$$E_{B,adv} = -e^{-i(kx - \omega t)} \qquad (C21)$$

for $t \leq T_2$, where

$$E_{A,adv} + E_{B,adv} = 0 \qquad (C22)$$

and where the advanced wave is zero for $t \geq T_2$.

So, we have net zero fields in the time intervals $t \leq T_1$ and $t \geq T_2$. In between, for $T_1 \leq t \leq T_2$, we have the nonzero net field

$$\begin{aligned} E_{net} &= E_A + E_B = E_{A,ret} + E_{B,adv} \\ &= \left[e^{i(kx - \omega t)} - e^{-i(kx - \omega t)} \right] \\ &= 2i\sin(kx - \omega t) \end{aligned} \qquad (C23)$$

where we recall that $E_{A,adv}$ and $E_{B,adv}$ are zero in this time interval.[3]

Note that the net field can be regarded as an electromagnetic wave traveling in the positive *x*-direction with increasing time *t*, or an electromagnetic wave of opposite sign moving in the negative *x*-direction with decreasing time *t*.

Thus, as observed in nature, an accelerated charged particle radiates and experiences radiation resistance when another charged particle is present. A lone charge, all by itself in the Universe, would not radiate. This is consistent with the view presented in chapter 3 that the acceleration of a body cannot be defined in the absence of other bodies (Leibniz-Mach principle). Radiation requires both a sender and a receiver. Furthermore, the sender is also a receiver, while the receiver is simultaneously a sender. The jiggling of charge A "causes" a "later" jiggling of charge B; but this jiggling of charge B "causes" the "earlier" jiggling of charge A. Radiation is thus an *interaction* between charges, a kind of handshake, and not a property of a single charge.

Thus, classical electrodynamics provides a consistent picture of non-self-acting charged particles, provided you include the impulses propagated backward in time as well as forward—exactly as the equations tell you that you must do! Even today, this fact is not appreciated. The great outstanding, unsolved problem of classical electrodynamics, which had been left hanging at the end of the nineteenth century, was in fact solved by Feynman and Wheeler by their allowing for electromagnetic effects to propagate backward as well as forward in time. Or, to put it another way, if you allow the equations to speak for themselves and do not impose on these equations the psychological prejudice of a single direction of time, you will avoid the paradox of infinite self-mass. I have argued that the so-called paradoxes of quantum mechanics result from the same stubborn unwillingness to believe what the equations tell us.[4]

NOTES

1. David Griffiths, *Introduction to Electrodynamics* (Upper Saddle River, NJ: Prentice-Hall, 1999), pp. 469–72.

2. S. S. Schweber, *QED and the Men Who Made It: Dyson, Feynman, Schwinger, and Tomonaga* (Princeton, NJ: Princeton University Press, 1994), p. 189.

3. This analysis forms the basis of John G. Cramer's "transactional interpretation" of quantum mechanics. See John G. Cramer, "The Transactional Interpretation of Quantum Mechanics," *Reviews of Modern Physics* 58 (1986): 647–88.

4. Victor J. Stenger, *Timeless Reality: Symmetry, Simplicity, and Multiple Universes* (Amherst, NY: Prometheus Books, 2000).

MATHEMATICAL SUPPLEMENT D

GAUGE INVARIANCE

GAUGE SYMMETRY

Let $q = \{q_o, q_1, q_2, q_3, \ldots q_{n-1}\}$ be the set of observables of a system and take them to be the coordinates of an n-dimensional vector q in q-space. Then a point in q-space, designated by the vector q, represents a particular set of measurements on a system. The generalized principle of covariance says that the laws of physics must be the same for any origin or orientation of q, that is, any choice of coordinate system.

Let us generalize space-time covariance further so that it also applies in the abstract space that contains our mathematical functions. We can define a vector ψ in that space as a set of functions of the observables of the system, $\psi = \{\psi_1(q), \psi_2(q), \psi_3(q), \ldots\}$. We will call this ψ-*space*. The state vectors of quantum mechanics are familiar examples of such abstract space vectors residing in *Hilbert space*; but, in general, ψ can represent any set of functions of observables that appears in the equations of physics.

The functions $\psi_1(q), \psi_2(q), \psi_3(q), \ldots$ comprise the projections of ψ on the coordinate axes in ψ-space. We assume that the following principle holds: *The models of physics cannot depend on the choice of coordinate system in ψ-space.* This principle is called *gauge invariance*. Another way to think of this is that the vector ψ is invariant under the transformation of coordinate systems, so that

$$\psi = \left\{\psi_1(q), \psi_2(q), \psi_3(q), \ldots\right\} = \left\{\psi_1'(q), \psi_2'(q), \psi_3'(q), \ldots\right\} \quad \text{(D1)}$$

where the first set of (unprimed) functions represents, say, the mathematical functions one theorist uses to describe the system, while the second set of (primed) functions are those of another theorist.

GAUGE TRANSFORMATIONS AND THEIR GENERATORS

To begin, let $\psi(q)$ be a complex function, that is, a two-dimensional vector in complex space, as shown in fig. 4.3 (p. 76). Let us perform a unitary transformation on ψ:

$$\psi' = U\psi \tag{D2}$$

where $U^\dagger U = 1$, so

$$\psi'^\dagger \psi' = \psi^\dagger U^\dagger U \psi = \psi^\dagger \psi \tag{D3}$$

This transformation does not change the magnitude of ψ,

$$\left|\psi'\right| = \left(\psi'^\dagger \psi'\right)^{\frac{1}{2}} = \left(\psi^\dagger \psi\right)^{\frac{1}{2}} = \left|\psi\right| \tag{D4}$$

That is, $|\psi|$ is *invariant* to the transformation, as required by gauge symmetry. We can write the operator U

$$U = \exp(i\theta) \tag{D5}$$

where $\theta^\dagger = q$, that is, θ is a *hermitian* operator (in this example, just a real number). Then,

$$\psi' = \exp(i\theta)\psi \tag{D6}$$

So, U changes the complex phase of ψ. It could be called a *phase transformation*, or simply a unitary transformation. However, with the amplifications of this idea that we will be discussing, the designation *gauge transformation* has become conventional. When θ is a constant we have a *global gauge transformation*. When θ is a not a constant, but a function of position and time, it is called a *local gauge transformation*.

Note also that the operation U corresponds to a rotation in the complex space of ψ. Later we will generalize these ideas to where ψ is a vector in higher dimensions and θ will be represented by a matrix. But this basic idea of a gauge transformation as analogous to a rotation in an abstract function space will be maintained, and gauge invariance will be viewed as an invariance under such rotations.

Let us write

$$\theta = -\varepsilon G \tag{D7}$$

where ε is an infinitesimal number and G is another operator. Then

$$U \approx 1 - i\varepsilon G \tag{D8}$$

where $G^\dagger = G$ is hermitian. G is called the *generator* of the transformation. Then,

$$\psi' \approx \psi - i\varepsilon G\psi \tag{D9}$$

Suppose we have a transformation that translates the q_μ-axis by an amount ε_μ. That is, the new coordinate $q_\mu' = q_\mu - \varepsilon_\mu$. Then, to first order in ε_μ,

$$
\begin{aligned}
\psi'(q_\mu') &= \psi(q_\mu - \varepsilon_\mu) \\
&\approx \psi(q_\mu) - \varepsilon_\mu \frac{\partial}{\partial q_\mu} \psi(q_\mu)
\end{aligned}
\tag{D10}
$$

It follows that the generator can be written

$$G_\mu = \frac{1}{i} \frac{\partial}{\partial q_\mu} \tag{D11}$$

Define

$$P_\mu \equiv \hbar G_\mu = \frac{\hbar}{i} \frac{\partial}{\partial q_\mu} \tag{D12}$$

where $\hbar$ is an *arbitrary constant* introduced only if you want the units of P_μ to be different from the reciprocal of the units of q_μ. The transformation operator can then be written

$$U = 1 - \frac{i}{\hbar} \varepsilon_\mu P_\mu \qquad (D13)$$

For example, suppose that $q_1 = x$, the x-coordinate of a particle. Then

$$P_1 \equiv P_x = \frac{\hbar}{i} \frac{\partial}{\partial x} \qquad (D14)$$

which we recognize as the quantum mechanical operator for the x-component of momentum. Note that this association was not assumed but derived, and that no connection with the physical momentum has yet been made. This just happens to be the mathematical form of the generator of a space translation. Similarly, we can take $q_2 = y$, $q_3 = z$ and obtain the generators P_x and P_y.

Of course, $\hbar$ will turn out to be the familiar *quantum of action* of quantum mechanics, $\hbar = h/2\pi$ where h is Planck's constant. Physicists often take $\hbar = 1$ in "natural units." We will leave $\hbar$ in our equations at this point to maintain familiarity; however, it should be recognized that the Planck constant, when expressed in dimensioned units, will turn out to be an arbitrary number defined only by that choice of units. No additional physical assumption about the "quantization of action" need be made and Planck's constant should not be viewed as a metric constant of nature. In particular, $\hbar$ cannot be zero. Once we have made the connection of (P_x, P_y, P_z) with the 3-momentum, quantization of action will already be in place.

We can also associate one of the variables, q_0, with the time t. In order to provide a connection with the fully relativistic treatment we will make later, let $q_0 = ict$, where c is, like $\hbar$, another arbitrary conversion factor. Then,

$$P_0 = \frac{\hbar}{i} \frac{\partial}{\partial q_0} = -\frac{\hbar}{c} \frac{\partial}{\partial t} \qquad (D15)$$

We can then define

$$H \equiv -iP_0 c = i\hbar \frac{\partial}{\partial t} \qquad (D16)$$

which we recognize as the quantum mechanical Hamiltonian (energy) operator. Note again that this familiar result was not assumed but derived from gauge transformations. No connection with the physical quantity energy has yet been made. This just happens to be the form of the generator of a time translation.

QUANTUM MECHANICS FROM GAUGE TRANSFORMATIONS

Suppose we have a complex function $\psi(x, y, z, t)$ that describes, in some still unspecified way, the state of a system. Let us make a gauge transformation of the time axis $t' = t - dt$

$$\psi'(t') = U\psi(t) = \exp\left(-\frac{i}{\hbar} H dt\right)\psi(t)$$

$$\approx \left(1 - \frac{i}{\hbar} H dt\right)\psi(t) = \psi(t) - dt\frac{\partial\psi}{\partial t}$$

(D17)

Then

$$H\psi = i\hbar\frac{\partial}{\partial t}\psi$$

(D18)

This is the time-dependent Schrödinger equation of quantum mechanics, where ψ is interpreted as the wave function.

At this point, then, we have the makings of quantum mechanics. That is, we have a mathematical model that looks like quantum mechanics, although we have not yet identified the operators H and P with the physical quantities energy and momentum. We have just noted that these are generators of time and space translations, respectively, which are themselves gauge transformations. We also have not yet specified the nature of the vector $\psi(q)$ except to say that it must be gauge invariant if it is to display point-of-view invariance.

In conventional quantum mechanics, $\psi(q)$ is defined so as to represent the state of a system and allow for the calculation of expected values of measurements. This idea is most elegantly expressed in the *bra* and *ket* notation for linear vectors introduced by Dirac. A vector in a linear vector space is written as either the *ket* $|a\rangle$ or its dual, *bra* $\langle a|$. The scalar product of two vectors $|a\rangle$ and $|b\rangle$ is written $\langle a|b\rangle$ and is, in general, a complex number where $\langle b|a\rangle = \langle a|b\rangle^*$.

In Dirac's formulation of quantum mechanics states are represented by linear vectors. For convenience, we often define the state vectors as orthonormal, that is,

$$\langle a|b\rangle = \delta_{ab} \tag{D19}$$

for all $|a\rangle$ and $|b\rangle$ that are state vectors.

The linearity postulate asserts that any state vector can be written as the superposition

$$|\psi\rangle = \sum_a |a\rangle\langle a|\psi\rangle \tag{D20}$$

This is also called the *superposition principle* and is responsible for much of the difference between quantum and classical mechanics, in particular, interference effects and so-called entangled states. As we saw in chapter 4, linearity and thus the superposition principle are required to maintain point-of-view invariance. That is, point-of-view invariance can be used as a motivation for the superposition principle, which, in conventional developments, is simply a hypothesis that almost seems to be pulled out of a hat.

The symbol $|a\rangle\langle a|$ can be viewed as an operator that projects $|\psi\rangle$ onto the $|a\rangle$ axis. The vectors $|a\rangle$ form a complete set,

$$\sum_a |a\rangle\langle a| = 1 \tag{D21}$$

so that $\langle\psi|\psi\rangle = 1$.

Consider a system with a single observable. When a measurement of that observable is made and found to have a value a, the system is said to be in a state $|a\rangle$. More generally, a will represent a set of measured observables $\{q\}$. We define an operator A such that

$$A|a\rangle = a|a\rangle \tag{D22}$$

That is, A is the linear vector operator that, when it operates on $|a\rangle$, gives the same vector scaled by the measurement value a. It is the operator representation of an observable. The above is called the *eigenvalue equation* of A, where $|a\rangle$ is the eigenvector or eigenstate of A corresponding to the eigenvalue a. The possible results of a measurement of the observable represented by A are given by the solutions of the eigenvalue equation. When the state of a system is an eigenstate of an observable, the measurement of that observable will always yield the eigenvalue corresponding to that state.

When the system is in an arbitrary state $|\psi\rangle$, then the expectation value of A, that is, the average or "expectation value" of an ensemble of measurements of A, is

$$\langle A \rangle = \langle \psi | A | \psi \rangle = \sum_a \sum_{a'} \langle \psi | a \rangle \langle a | A | a' \rangle \langle a' | \psi \rangle$$
$$= \sum_a |\langle \psi | a \rangle|^2 a \tag{D23}$$

Thus, $|\langle \psi | a \rangle|^2$ is interpreted as the probability for $|\psi\rangle$ to be found in the state $|a\rangle$.

We can also express this in more familiar wave-function language, although it is to be noted that Dirac disdained the use of the term. The wave function is defined as the scalar product

$$\psi(q) = \langle q | \psi \rangle \tag{D24}$$

where $|q\rangle$ are the eigenstates of the space-time coordinates of the particles of the system. Momentum-space wave functions are also often used.

For example, the basis states are frequently chosen to be $|x\rangle$, $|y\rangle$, and $|z\rangle$, where the observables are all the possible coordinates of a particle, that is, all the eigenstates of the eigenvalue equation

$$X | x \rangle = x | x \rangle \tag{D25}$$

with similar equations for Y and Z. The quantity $\psi(x) = \langle x | \psi \rangle$ is, for historical reasons, called the *wave function*, although it usually has nothing to do with waves. Since x is conventionally regarded as a continuous variable, ψ-space is infinite dimensional! That is, $|x\rangle$ is not one axis but an infinite number of axes, one for every real number x. Even if we assume that x is discrete, say, in units of the Planck length, and space is finite, we still have an awfully large number of dimensions.

More generally, the eigenstates $|q\rangle$ are the basis states of a particular, arbitrary representation, like the unit vectors **i**, **j**, and **k** of the Cartesian coordinate axes x, y, z. $\psi(q)$ is the projection of $|\psi\rangle$ on $|q\rangle$.

We can also represent ψ and q as 1-dimensional matrices. Then $\psi(q)$ is the matrix scalar product

$$\psi(q) = \sum_i \psi_i^\dagger q_i \tag{D26}$$

where $\psi_i^\dagger$ is a row matrix.

In this representation, the observable A is a square matrix, and its expectation value

$$\langle A \rangle = \sum_{i,j} \psi_i^\dagger A_{ij} \psi_j \tag{D27}$$

is simply given by the matrix equation,

$$\langle A \rangle = \psi^\dagger A \psi \tag{D28}$$

Thus, the gauge transformation can be written as a matrix equation

$$\psi' = U\psi = \exp(i\theta)\psi \tag{D29}$$

where U and θ, and the corresponding generators, are square matrices.

Let us return to bra and ket notation, where we write the gauge transformation

$$|\psi'\rangle = U|\psi\rangle \tag{D30}$$

The expectation value of an operator A

$$\langle A \rangle = \langle \psi | A | \psi \rangle = \langle \psi' | U^\dagger A U | \psi' \rangle = \langle \psi' | A' | \psi' \rangle \tag{D31}$$

is gauge invariant, where the transformed operator is

$$A' = U^\dagger A U \tag{D32}$$

Let us look again at time evolution. There are two approaches. In the *Schrödinger picture*, the state vector varies with time while the operators stay fixed.

$$
\begin{aligned}
A(t) &= A(0) \\
|\psi(t)\rangle &= U|\psi(0)\rangle \\
\langle A(0)\rangle &= \langle\psi(0)|A(0)|\psi(0)\rangle \\
\langle A(t)\rangle &= \langle\psi(t)|A(0)|\psi(t)\rangle \\
&= \langle\psi(0)|U^\dagger A(0)U|\psi(0)\rangle
\end{aligned}
\tag{D33}
$$

In the *Heisenberg picture*, the state vectors remain fixed while the operators evolve with time.

$$
\begin{aligned}
A(t) &= U^\dagger A(0)U \\
\psi(t) &= \psi(0) \\
\langle A(0)\rangle &= \langle\psi(0)|A(0)|\psi(0)\rangle \\
\langle A(t)\rangle &= \langle\psi(0)|A(t)|\psi(0)\rangle \\
&= \langle\psi(0)|U^\dagger A(0)U|\psi(0)\rangle
\end{aligned}
\tag{D34}
$$

Notice the same result in either case. We can write

$$
\begin{aligned}
U &= \exp\left(-\frac{i}{\hbar}Hdt\right) \rightarrow \left(1-\frac{i}{\hbar}Hdt\right) \\
A(t+dt) &= U^\dagger A(t)U \\
&= \left(1+\frac{i}{\hbar}Hdt\right)A(t)\left(1-\frac{i}{\hbar}Hdt\right) \\
&= A(t) - \frac{i}{\hbar}\left[A,H\right]dt = A(t) + dt\frac{\partial A}{\partial t} \\
\Rightarrow\quad \frac{\partial A}{\partial t} &= -\frac{i}{\hbar}\left[A,H\right]
\end{aligned}
\tag{D35}
$$

So, the time rate of change of an operator is

$$\frac{dA}{dt} = \frac{\partial A}{\partial t} + \sum_{j \neq 0} \frac{\partial A}{\partial q_j} \frac{dq_j}{dt}$$

$$= -\frac{i}{\hbar} [A, H] + \sum_{j=1}^{n} \frac{\partial A}{\partial q_j} \frac{dq_j}{dt}$$

(D36)

Next, let us move to gauge transformations involving the non-temporal variables of a system. Consider the case where $A = P_j$. Then,

$$\frac{dP_j}{dt} = -\frac{i}{\hbar} [P_j, H] + \sum_{k=1}^{n} \frac{\partial P_j}{\partial q_k} \frac{dq_k}{dt}$$

(D37)

Consider the transformation of these nontemporal variables. Let $q_k' = q_k - \varepsilon_k$, which corresponds to translating the q_k-axis by an infinitesimal amount ε_k. As we saw in equation (D13), the transformation operator is

$$U = 1 - \frac{i}{\hbar} P_k \varepsilon_k$$

(D38)

Thus,

$$\left| \psi'(q_k') \right\rangle = \left| \psi(q_k - \varepsilon_k) \right\rangle$$

$$= \left| \psi(q_k) \right\rangle - \frac{i}{\hbar} P_k \varepsilon_k \left| \psi(q_k) \right\rangle$$

(D39)

So

$$A' = U^\dagger A U = \left(1 + \frac{i}{\hbar} P_k \varepsilon_k \right) A \left(1 - \frac{i}{\hbar} P_k \varepsilon_k \right)$$

$$= 1 - \frac{i}{\hbar} \varepsilon_k [A, P_k]$$

(D40)

and

$$\frac{\partial A}{\partial q_k} = -\frac{i}{\hbar} [A, P_k]$$

(D41)

From the differential form of the operators P_k,

$$[P_j, P_k] = 0$$

(D42)

and so

$$\frac{\partial P_j}{\partial q_k} = 0 \tag{D43}$$

Recall that

$$\frac{dP_k}{dt} = -\frac{i}{\hbar}\left[P_k, H\right] + \sum_j \frac{\partial P_k}{\partial q_j}\frac{dq_j}{dt} \tag{D44}$$

The summed terms are all zero, so

$$\frac{dP_k}{dt} = -\frac{i}{\hbar}\left[P_k, H\right] \tag{D45}$$

We can also think of q_k as an operator, so

$$\frac{\partial q_k}{\partial q_k} = 1 = -\frac{i}{\hbar}\left[q_k, P_k\right] \tag{D46}$$

or

$$\left[q_k, P_k\right] = i\hbar \tag{D47}$$

This can also be seen from

$$\left[q_k, P_k\right]\psi = \left[q_k, \frac{\hbar}{i}\frac{\partial}{\partial q_k}\right]\psi$$

$$= \frac{\hbar}{i}q_k\frac{\partial\psi}{\partial q_k} - \frac{\hbar}{i}\frac{\partial}{\partial q_k}q_k\psi = i\hbar\psi \tag{D48}$$

For example,

$$\left[x, P_x\right] = i\hbar \tag{D49}$$

the familiar quantum mechanical commutation relation.

Now we can also write

$$\frac{\partial H}{\partial q_k} = \frac{i}{\hbar}\left[H, P_k\right] \tag{D50}$$

Thus,

$$\frac{dP_k}{dt} = -\frac{\partial H}{\partial q_k} \tag{D51}$$

which is the operator version of one of Hamilton's classical equations of motion and another way of writing Newton's second law of motion. Here we see that we have developed another profound concept from gauge invariance alone. When the Hamiltonian of a system does not depend on a particular variable, then the observable corresponding to the generator of the gauge transformation of that variable is conserved. This is a generalized version of *Noether's theorem* for dimensions other than space and time. Note that by including the space-time coordinates as part of our set of abstract coordinates we unite all the conservation principles under the umbrella of gauge symmetry.

THE UNCERTAINTY PRINCIPLE

As we found above, certain pairs of operators do not mutually commute. Consider two such operators, where

$$\left[A, B\right] \neq 0 \tag{D52}$$

Let

$$\Delta A \equiv A - \left\langle A \right\rangle \tag{D53}$$

where $\left\langle A \right\rangle$ is the mean value of a set of measurements of A. The dispersion (or variance) of A is defined as

$$\left\langle \left(\Delta A\right)^2 \right\rangle = \left\langle \left(A^2\right) - 2A\left\langle A \right\rangle + \left\langle A \right\rangle^2 \right\rangle = \left\langle A^2 \right\rangle - \left\langle A \right\rangle^2 \tag{D54}$$

with a similar definition for ΔB. In advanced quantum mechanics textbooks[1] you will find derivations of the *Schwarz inequality:*

$$\left\langle \alpha | \alpha \right\rangle \left\langle \beta | \beta \right\rangle \geq \left| \left\langle \alpha | \beta \right\rangle \right|^2 \tag{D55}$$

from which it can be shown that

$$\left\langle \left(\Delta A \right)^2 \right\rangle \left\langle \left(\Delta B \right)^2 \right\rangle \geq \frac{1}{4} \left| \left\langle \left[A, B \right] \right\rangle \right|^2 \tag{D56}$$

which is the generalized Heisenberg uncertainty principle. For example, as we saw above,

$$\left[x, P_x \right] = i\hbar \tag{D57}$$

from which it follows that

$$\Delta x \Delta P_x \geq \frac{\hbar}{2} \tag{D58}$$

ROTATION AND ANGULAR MOMENTUM

The variables (q_1, q_2, q_3) can be identified with the coordinates (x, y, z) of a particle, and the corresponding momentum components are the generators of translations of these coordinates. (In this formulation, nothing prevents other particles from being included with their space-time variables associated with other sets of four q's; note that by having each particle carry its own time coordinate we can maintain a fully relativistic scheme.) These coordinates may also be angular variables and their conjugate momenta may be the corresponding angular momenta. These angular momenta will be conserved when the Hamiltonian is invariant to the gauge transformations that correspond to rotations by the corresponding angles about the spatial axes. For example, if we take $(q_1, q_2, q_3) = (\phi_x, \phi_y, \phi_z)$, where ϕ_x is the angle of rotation about the x-axis, and so on, then the generators of the rotations about these axes will be the angular momentum components (L_x, L_y, L_z). Rotational invariance about any of these axes will lead to conservation of angular momentum about that axis.

Let us look at rotations in familiar 3-dimensional space. Suppose we have a vector $\mathbf{V} = (V_x, V_y)$ in the x-y plane. Let is rotate it counterclockwise about the z-axis by an angle ϕ. We can write the transformation as a matrix equation

$$\begin{pmatrix} V'_x \\ V'_y \end{pmatrix} = \begin{pmatrix} \cos\phi & -\sin\phi \\ \sin\phi & \cos\phi \end{pmatrix} \begin{pmatrix} V_x \\ V_y \end{pmatrix} \tag{D59}$$

Specifically, let us consider an infinitesimal rotation of the position vector $\mathbf{r} = (x, y)$ by $d\phi$ about the z-axis. From above,

$$\begin{pmatrix} x' \\ y' \end{pmatrix} = \begin{pmatrix} 1 & -d\phi \\ d\phi & 1 \end{pmatrix} \begin{pmatrix} x \\ y \end{pmatrix} = \begin{pmatrix} x - yd\phi \\ y + xd\phi \end{pmatrix} \tag{D60}$$

And so,

$$dx = -yd\phi \tag{D61}$$

and

$$dy = xd\phi \tag{D62}$$

For any function $f(x, y)$,

$$f(x + dx, y + dy) = f(x, y) + dx\frac{\partial f}{\partial x} + dy\frac{\partial f}{\partial y} \tag{D63}$$

to first order. Or we can write (reusing the function symbol f)

$$f(\phi + d\phi) = f(\phi) - yd\phi\frac{\partial f}{\partial x} + xd\phi\frac{\partial f}{\partial y} \tag{D64}$$

from which we determine that the generator of a rotation about z is

$$G = -i\left(x\frac{\partial}{\partial y} - y\frac{\partial}{\partial x} \right) = xP_y - yP_x = L_z \tag{D65}$$

which is also the angular momentum about z. Similarly,

$$L_x = yP_z - zP_y \tag{D66}$$

and

$$L_y = zP_x - xP_z \tag{D67}$$

This result can be generalized as follows. If you have a function that depends on a spatial position vector $\mathbf{r} = (x, y, z)$, and you rotate that position vector by an angle θ about an arbitrary axis, then that function transforms as

$$f'(\mathbf{r}) = \exp(i\mathbf{L} \bullet \boldsymbol{\theta}) f(\mathbf{r}) \tag{D68}$$

where the direction of the axial vector $\boldsymbol{\theta}$ is the direction of the axis of rotation. Once again this has the form of a gauge transformation, or phase transformation of f, where the transformation operator is

$$U = \exp(i\mathbf{L} \bullet \boldsymbol{\theta}) \tag{D69}$$

From the previous commutation rules one can show that the generators L_x, L_y, and L_z do not mutually commute. Rather,

$$\left[L_x, L_y \right] = i\hbar L_z \tag{D70}$$

and cyclic permutations of x, y, and z. Thus the order of successive rotations is important. Note that, from (D56),

$$\Delta L_x \Delta L_y = \frac{1}{2} \left| \left\langle \left[L_x, L_y \right] \right\rangle \right| = \frac{\hbar}{2} \left\langle L_z \right\rangle \tag{D71}$$

Most quantum mechanics textbooks contain the proof of the following result, although it is not always stated so generally: Any vector operator **J** whose components obey the angular momentum commutation rules,

$$\left[J_x, J_y \right] = i\hbar J_z \tag{D72}$$

and cyclic permutations will have the following eigenvalue equations

$$J^2 \left| j, m \right\rangle = j(j+1)\hbar^2 \left| j, m \right\rangle \tag{D73}$$

where $J^2 = J_x^2 + J_y^2 + J_z^2$ is the square of the magnitude of **J**, and

$$J_z \left| j, m \right\rangle = m\hbar \left| j, m \right\rangle \tag{D74}$$

where m goes from $-j$ to $+j$ in steps of one: $m = -j, -j+1, \ldots, j-1, j$. Furthermore, $2j$ is an integer. This implies that j is an integer (including zero) or a half-integer. In particular, note that the half-integer nature of the spins of fermions is a consequence of angular momentum being the generator of rotations.

ROTATION AND GAUGE TRANSFORMATIONS

We have already noted that the gauge transformation is like a rotation in the complex space of a vector $|\psi\rangle$, which in quantum mechanics represents the state of the system. The basis states in ψ-space are the eigenstates of the observables of the system. For example, those basis states may be $|x\rangle$, the eigenstates of the position of a particle along the x-axis—meaning one axis for every possible value of x. If the particle is an electron, then ψ-space must also include the basis states $|+\frac{1}{2}\rangle$ and $|-\frac{1}{2}\rangle$ that are the eigenstates of the z-component of the spin of the electron. Even though spatial coordinates are more familiar than spins, two-dimensional spin subspace is a lot easier to visualize than the infinite-dimensional subspace of spatial coordinate eigenstates.

In the two-dimensional subspace spanned by the spin state vector of an electron, the basis states $|+\frac{1}{2}\rangle$ and $|-\frac{1}{2}\rangle$ can be thought of as analogous to the unit vectors $\mathbf{i}$ and $\mathbf{j}$ in the more familiar two-dimensional subspace (x, y). The spin state $|\psi\rangle$ is in general a two-dimensional vector oriented at some arbitrary angle. The basis vectors define two possible orientations of the spin angular momentum vector $\mathbf{S}$ in familiar three-dimensional space, one along the z-axis and the other along the opposite. (The choice of z-axis here is an arbitrary convention.) Thus, for example, if $\mathbf{S}$ points originally along the z-axis, a rotation of $180°$ will take it to point along $-z$. However, note that a rotation in ψ-space of only $90°$ takes the spin state from $|+\frac{1}{2}\rangle$ to $|-\frac{1}{2}\rangle$, as illustrated in Fig. D4.1. This implies that the unitary transformation matrix in this case is

$$U = \exp\left(i\frac{\theta}{2}\right)I \tag{D75}$$

where I is the unit 2×2 matrix.

More generally,

$$U = \exp\left(i\frac{\boldsymbol{\sigma} \bullet \boldsymbol{\theta}}{2}\right) \tag{D76}$$

where the unit matrix I is understood, the axial vector $\boldsymbol{\theta}$ points in the direction around which we rotate, and $\boldsymbol{\sigma}$ is the Pauli spin vector (that is,

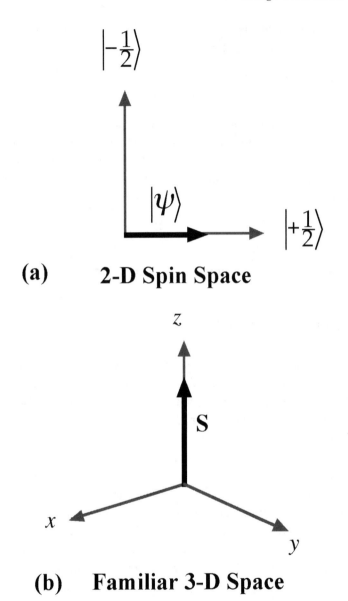

(a)   2-D Spin Space

(b)   Familiar 3-D Space

Fig. D4.1. Comparison of two-dimensional spin space with familiar three-dimensional space. The spin vector **S** points in the z-direction, "up." The state vector $|\psi\rangle$ points in the direction of the corresponding eigenstate. If **S** is rotated by 180° so it points "down," the state vector is rotated by 90° in spin space.

a vector in coordinate space and a matrix in spin space), whose components are conventionally written

$$\sigma_x = \begin{pmatrix} 0 & 1 \\ 1 & 0 \end{pmatrix} \quad \sigma_y = \begin{pmatrix} 0 & -i \\ i & 0 \end{pmatrix} \quad \sigma_z = \begin{pmatrix} 1 & 0 \\ 0 & -1 \end{pmatrix} \tag{D77}$$

We see that U again has the form of a gauge transformation. The generator of the gauge transformation in the spin vector subspace of a spin-1/2 particle is the spin angular momentum operator (in units of $\hbar$), $\mathbf{S} = \boldsymbol{\sigma}/2$. We could also have obtained this result from our previous proof that the gauge transformation for a rotation in 3-space is, as given by (D69),

$$U = \exp(i\mathbf{L} \bullet \boldsymbol{\theta}) \tag{D78}$$

where L is the angular momentum. Here $\mathbf{L} = \mathbf{S} = \boldsymbol{\sigma}/2$.

SPECIAL RELATIVITY

Now we are ready to inject some familiar physics into the mix. It turns out to be easiest to do this within the framework of special relativity. But note that, as was the case for quantum mechanics, the usual starting axioms will not be asserted. Rather they will be derived from the assumptions of space-time covariance and gauge invariance.

Let us consider the first four variables (q_0, q_1, q_2, q_3) of our set $\{q\}$ that we have arbitrarily set to $(x_0, x_1, x_2, x_3) = (ict, x, y, z)$, where t is the time and where (x, y, z) are the spatial coordinates of an event. The constant c is simply a factor that converts units of time into units of distance. It will turn out to be the invariant speed of light in a vacuum, but that is not being assumed at this point. Also, the assumption that q_0 is an imaginary number is not necessary; it just makes things easier to work out.

Let $x' = (x'_0, x'_1, x'_2, x'_3)$ be the position of the event in a reference frame moving at a speed $v = \beta c$ along the z-axis with respect to the reference frame x, where

$$x'_\mu = L^v_\mu x_v \tag{D79}$$

and the convention is used in which repeated Greek indices are summed from 0 to 3. As is shown in many textbooks, the proper distance will be invariant if L_μ^ν is the Lorentz transformation operator

$$L_\mu^\nu = \begin{pmatrix} 1 & 0 & 0 & 0 \\ 0 & 1 & 0 & 0 \\ 0 & 0 & \cos\psi & \sin\psi \\ 0 & 0 & -\sin\psi & \cos\psi \end{pmatrix} \tag{D80}$$

where $\cos\psi = \gamma$, $\sin\psi = i\beta\gamma$, and $\gamma = (1 - \beta^2)^{1/2}$. By writing it this way, we see that the Lorentz transformation between reference frames moving at constant velocity with respect to one another along their respective z-axes is equivalent to a rotation by an angle ψ in the (x_1, x_2)-plane. That is, Lorentz invariance is analogous to rotational invariance in 3-space.

The complex angle ψ is a mathematical artifact, taking the zeroth component of the 4-vector to be an imaginary number and time a real number. We could make ψ real by using a non-Euclidean metric.

We have seen that the generators of space-time translation form a 4-component set:

$$P = \left(P_0, P_1, P_2, P_3\right) = \left(i\frac{H}{c}, P_x, P_y, P_z\right) \tag{D81}$$

where we recall that c is just a unit-conversion constant. Quantum mechanically,

$$P_k\left|p_k\right\rangle = p_k\left|p_k\right\rangle \tag{D82}$$

where p_k is the eigenvalue of P_k when the system is in a state given by the eigenvector $\left|p_k\right\rangle$. Similarly,

$$H\left|E\right\rangle = E\left|E\right\rangle \tag{D83}$$

Let us work with these eigenvalues—which have still not been identified with familiar physical energy and momentum! But that's coming up fast now. Write

$$p = \left(p_0, p_1, p_2, p_3\right) = \left(i\frac{E}{c}, p_x, p_y, p_z\right) \tag{D84}$$

The squared length of the 4-vector

$$p_\mu p^\mu = p'_\mu p'^\mu \equiv -m^2 c^2 \tag{D85}$$

is invariant to rotations in 4-space. The invariant quantity m is called the *mass* of the particle. Note that the length of the 4-momentum vector is (in the metric I have chosen to use)

$$\left(p_\mu p^\mu\right)^{1/2} = imc \tag{D86}$$

Defining 4-momentum in this way guarantees the invariance in the important result of an earlier section, namely, the classical Hamilton equation of motion

$$\frac{dP_k}{dt} = -\frac{\partial H}{\partial q_k} \tag{D87}$$

This definition allows us to connect the operator P_k with the operationally defined momentum p_k and the operator H with the operationally defined energy E.

Working with the operationally defined quantities, we can write (using boldface type for familiar 3-dimensional spatial vectors)

$$d\mathbf{p} \bullet d\mathbf{r} = -dE dt \tag{D88}$$

Or, in terms of 4-vectors,

$$dp^\mu dx_\mu = 0 \tag{D89}$$

which is Lorentz invariant.

Suppose we have a particle of mass m. Let (x', y', z') be the coordinate axes in the reference frame in which the particle is at rest, $|\mathbf{p'}| = 0$. Then its energy in that reference frame is

$$E' = mc^2 \tag{D90}$$

which is the *rest energy*.

Next, let us look at the particle in another reference frame (x, y, z) in which the particle is moving along the z-axis at a constant speed v. Then, from the Lorentz transformation, the 3-momentum of the particle in that reference frame will be

$$p_z = \gamma\left(p'_z + \frac{\beta}{c}E'\right) = \gamma(0 + \beta mc) \tag{D91}$$

We can write this in vector form as

$$\mathbf{p} = \gamma m\mathbf{v} \tag{D92}$$

We note that $\mathbf{p} \to m\mathbf{v}$ when $v \ll c$. So, we have (finally) derived the well-known relationship between momentum and velocity. Nowhere previously was it assumed that $\mathbf{p} = m\mathbf{v}$.

The energy of the particle in the same reference frame is

$$E = \gamma\left(E' + \beta p'_z\right) = \gamma mc^2 \tag{D93}$$

Note that, in general, the velocity of a particle is

$$\mathbf{v} = \frac{\mathbf{p}c^2}{E} \to \frac{\mathbf{p}}{m} \tag{D94}$$

when $v \ll c$ since, in that case, $E = mc^2$. We can also show that for all v,

$$E = \left(\left|\mathbf{p}c\right|^2 + m^2c^4\right)^{1/2} \tag{D95}$$

This is a "free particle" since

$$\mathbf{F} = \frac{d\mathbf{p}}{dt} = -\nabla E = 0 \tag{D96}$$

More generally we can write

$$E = mc^2 + T + V(\mathbf{r}) \tag{D97}$$

where mc^2 is the rest energy. The quantity

$$T = \left(\left|\mathbf{p}c\right|^2 + m^2c^4\right)^{1/2} - mc^2 \tag{D98}$$

is the *kinetic energy*, or energy of motion, where

$$T \rightarrow \frac{1}{2}mv^2 \tag{D99}$$

when $v \ll c$. $V(\mathbf{r})$ is the *potential energy*. The force on the particle is then

$$\mathbf{F} = -\nabla V \tag{D100}$$

We are now in a position to interpret the meaning of c, which was introduced originally as a simple conversion factor. Suppose we have a particle of zero mass and 3-momentum of magnitude $|\mathbf{p}|$. Then, the energy of that particle will be

$$E = |\mathbf{p}|c \tag{D101}$$

and the speed will be

$$v = \frac{pc^2}{E} \rightarrow \frac{pc^2}{pc} = c \tag{D102}$$

Thus c is the speed of a zero-mass particle, sometimes called the *speed of light*. Since c is the same constant in all references frames, the invariance of the speed of light, one of the axioms of special relativity, is thus seen to follow from 4-space rotational symmetry.

So we have now shown that the generators of translations along the four axes of space-time are the components of the 4-momentum, which includes energy in the zeroth component and 3-momentum in the other components. These have their familiar connections with the quantities of classical physics. Mass is introduced as a Lorentz-invariant quantity that is proportional to the length of the 4-momentum vector. The conversion factor c is shown to be, as expected, the Lorentz-invariant speed of light in a vacuum.

CLASSICAL MECHANICS

Except for specific laws of force for gravity and electromagnetism, all of classical mechanics can now be inferred from the above discussion. Conservation of energy, linear momentum, and angular momentum follow from global gauge invariance in space-time. Newton's first and third laws of motion follow from momentum conservation. Newton's second law basically defines the force on a body as the time rate of change of momentum,

$$\mathbf{F} = \frac{d\mathbf{p}}{dt} \tag{D103}$$

Above we saw that, for the operators $\mathbf{P}$ and H,

$$\frac{d\mathbf{P}}{dt} = -\nabla H \tag{D104}$$

The classical observables will correspond to the eigenvalues of these and so

$$\frac{d\mathbf{p}}{dt} = -\nabla E \tag{D105}$$

If $E = T + V$ and T does not depend explicitly on spatial position,

$$\mathbf{F} = \frac{d\mathbf{p}}{dt} = -\nabla V \tag{D106}$$

as in the previous section. The more generalized and advanced formulations of classical mechanics, such as Lagrange's and Hamilton's equations of motion, can be developed in the usual way.

ELECTROMAGNETISM

In the following sections we will switch to the conventions used in elementary particle physics so that the resulting equations agree with the textbooks at that level. We have already seen that $\hbar$ and c are arbitrary conversion factors, so we will work in units where $\hbar = c = 1$. Furthermore, we will use a non-Euclidean (but still geometrically flat) metric in defining our 4-vectors:

$$\eta^{\nu}_{\mu} = \begin{pmatrix} 1 & 0 & 0 & 0 \\ 0 & -1 & 0 & 0 \\ 0 & 0 & -1 & 0 \\ 0 & 0 & 0 & -1 \end{pmatrix} \tag{D107}$$

where the position 4-vector is $x = (t, x, y, z)$, and where we reuse x; the momentum 4-vector is $p = (E, p_x, p_y, p_z)$, and

$$p_\mu \eta^\mu_\nu p^\nu = E^2 - |\mathbf{p}|^2 = m^2 \tag{D108}$$

This choice of metric has the advantages of enabling us to directly identify the mass with the invariant length of the 4-momentum vector and eliminating the need for an imaginary zeroth component.

In quantum mechanics, the state of a free particle is an eigenstate of energy and momentum. Consider the 4-momentum eigenvalue equation for a spinless particle (spin can be included, but this is sufficient for present purposes)

$$i\partial_\mu \phi = p_\mu \phi \tag{D109}$$

where we now use the convention $\partial_\mu \phi \equiv \dfrac{\partial \phi}{\partial x_\mu}$.

The quantity ϕ is the eigenfunction $\phi(x) = \langle x | p_\mu \rangle$ and can be thought of as having two abstract dimensions, its real and imaginary parts. If we rotate the axis in this space by an angle θ, we have the gauge transformation,

$$\phi' = \exp\left(i\theta\right)\phi \tag{D110}$$

The eigenvalue equation is unchanged, provided that θ is independent of the space-time position x. This is the type of gauge invariance we have already considered, what we call global gauge invariance. The generator of the transformation, θ, is conserved. Below we will identify θ with the negative of the charge of the particle.

Now suppose that θ depends on the space-time position x. In this case, we do a *local* gauge transformation so that

$$\partial_\mu \phi' = \exp\left(i\theta(x)\right)\partial_\mu \phi + \exp\left(i\theta(x)\right)\phi i\left(\partial_\mu \theta(x)\right) \tag{D111}$$

That is, the eigenvalue equation is not invariant to this operation. Let us define a new operator, the *covariant derivative*

$$\mathcal{D}_\mu = \partial_\mu + iqA_\mu \qquad \text{(D112)}$$

where q is a constant and A_μ transforms as

$$A'_\mu = A_\mu + \partial_\mu \xi(x) \qquad \text{(D113)}$$

where

$$\theta(x) = -q\xi(x) \qquad \text{(D114)}$$

Then,

$$
\begin{aligned}
\mathcal{D}'_\mu \phi' &= \left(\partial_\mu + iqA'_\mu\right)\phi \\
&= \left[\partial_\mu + iqA_\mu + iq\left(\partial_\mu \xi\right)\right]\exp(i\theta)\phi \\
&= \exp(i\theta)\left[\partial_\mu + iqA_\mu + iq\left(\partial_\mu \xi\right)\right]\phi - \exp(i\theta)\phi\left[iq\left(\partial_\mu \xi\right)\right] \\
&= \exp(i\theta)\left(\partial_\mu + iqA_\mu\right)\phi \\
&= \exp(i\theta)\mathcal{D}_\mu\phi
\end{aligned}
\qquad \text{(D115)}
$$

Recall that the operator P_μ associated with the relativistic 4-momentum is

$$P_\mu = -i\partial_\mu \qquad \text{(D116)}$$

Let us define, analogously,

$$\mathcal{P}_\mu = -i\mathcal{D}_\mu \qquad \text{(D117)}$$

Writing

$$\mathcal{P}_\mu = P_\mu + qA_\mu \qquad \text{(D118)}$$

we see that this operator $\mathcal{P}_\mu$ is precisely the canonical 4-momentum in classical mechanics for a particle of charge q interacting with an electromagnetic field described by the 4-vector potential $A_\mu = (A_o, \mathbf{A})$, where

$A_0 = V/c$ in terms of the scalar potential V and $\mathbf{A}$ is the 3-vector potential. We will justify this connection further below. Since $\theta(x) = -q\xi(x)$, q is conserved when $\xi(x)$ is a constant. Also, note that for neutral particles, $q = 0$ and no new fields need to be introduced to preserve gauge invariance in that case.

Also note that q is the generator of a rotation in a two-dimensional space, which is mathematically an angular momentum. Thus global gauge invariance in this space will result in q being quantized. That is, charge quantization is yet another consequence of point-of-view invariance.

In quantum mechanics, the canonical momentum must be quantized in place of the mechanical momentum in the presence of an electromagnetic field. For example, the Schrödinger equation for a nonrelativistic particle of mass m and charge q in an electromagnetic field described by the 3-vector potential $\mathbf{A}$ and the scalar potential V is

$$\left(\left| \mathbf{P} - q\mathbf{A} \right|^2 + qV \right)\psi = \left(\left| -i\hbar\nabla - q\mathbf{A} \right|^2 + qV \right)\psi$$
$$= i\hbar\frac{\partial\psi}{\partial t}$$

(D119)

In quantum field theory, the basic quantity from which calculations proceed is the *Lagrangian density*, $\mathcal{L}$. The Klein-Gordon Lagrangian density for the field of a spinless particle of mass m is

$$\mathcal{L} = -\tfrac{1}{2}\partial^{\mu}\partial_{\mu}\phi + \tfrac{1}{2}m^2\phi^2$$

(D120)

Just as we can obtain the equation of motion of a particle from the Lagrangian by using Lagrange's equation, we can obtain the equation of motion of a field from the Lagrangian density and Lagrange's density equation

$$\partial^{\mu}\left(\frac{\partial\mathcal{L}}{\partial\partial_{\mu}\phi} \right) - \frac{\partial\mathcal{L}}{\partial\phi} = 0$$

(D121)

In this case the equation of motion is the Klein-Gordon equation,

$$\partial^{\mu}\partial_{\mu}\phi + m^2\phi = 0$$

(D122)

Now, the Klein-Gordon Lagrangian density is not locally gauge invariant. However, it becomes so if we write it

$$\mathcal{L} = -\tfrac{1}{2}\mathcal{D}^{\dagger\mu}\mathcal{D}_{\mu}\phi + \tfrac{1}{2}m^2\phi^2 \tag{D123}$$

The corresponding Klein-Gordon equation, the relativistic analogue of the Schrödinger equation for spinless particles, becomes

$$\mathcal{D}^{\dagger\mu}\mathcal{D}_{\mu}\phi + m^2\phi = 0 \tag{D124}$$

Spin-1/2 particles of mass m are described by the Dirac Lagrangian density, which similarly can be made gauge invariant by writing it, using conventional notation,

$$\mathcal{L} = i\bar{\psi}\gamma^{\mu}\mathcal{D}_{\mu}\psi - m\bar{\psi}\psi \tag{D125}$$

where the γ_{μ} are a set of 4×4 matrices (see any textbook on relativistic quantum mechanics). The corresponding Dirac equation

$$i\gamma^{\mu}\mathcal{D}_{\mu}\psi - m\psi = 0 \tag{D126}$$

also is gauge invariant.

A spin-1 particle of mass m_A is described by the Proca Lagrangian density

$$\mathcal{L} = -\frac{1}{16\pi}F^{\mu\nu}F_{\mu\nu} + \frac{1}{8\pi}m_A^2 A^{\mu}A_{\mu} \tag{D127}$$

where

$$F_{\mu\nu} = \partial_{\mu}A_{\nu} - \partial_{\nu}A_{\mu} \tag{D128}$$

Standard international units are used here. Some more advanced texts work in Heaviside-Lorentz units in which the above expression for $\mathcal{L}$ is multiplied by 4π. The first term in $\mathcal{L}$ is gauge invariant while the second term is *not* unless we set $m_A = 0$. This leads to the deeply important result that particles with spin-1 whose Lagrangians are locally gauge invariant are necessarily massless. The photon is one such particle. However, other spin-1 fundamental particles exist with nonzero masses. These masses result from spontaneous broken symmetry.

In any case, the existence of a vector field A_μ associated with a massless spin-1 particle is implied by the assumption of local gauge invariance. It is a field introduced to maintain local gauge invariance. That field can be identified with the classical electromagnetic fields **E** and **B,** and the particle with the photon. That is, the photon is the quantum of the field A_μ, which itself is associated with the classical 4-vector electromagnetic potential.

To see the classical connection, note that, from (D113),

$$A'_k = A_k + \partial_k \xi(x) \tag{D129}$$

where $k = 1, 2, 3$, or, in familiar 3-vector notation

$$\mathbf{A}' = \mathbf{A} + \nabla \xi \tag{D130}$$

This requires that the 3-vector

$$\begin{aligned} \mathbf{B}' = \nabla \times \mathbf{A}' = \nabla \times \mathbf{A} + \nabla \times \nabla \xi \\ = \nabla \times \mathbf{A} = \mathbf{B} \end{aligned} \tag{D131}$$

is locally gauge invariant. Furthermore,

$$\nabla \bullet \mathbf{B} = \nabla \bullet \left(\nabla \times \mathbf{A} \right) = 0 \tag{D132}$$

Thus, **B** may be interpreted as the familiar classical magnetic field 3-vector; the above equation is Gauss's law of magnetism, one of Maxwell's equations.

The zeroth component of the 4-vector potential

$$A'_o = A_o + \frac{\partial \xi}{\partial x_o} \tag{D133}$$

can be written

$$V' = V + \frac{\partial \xi}{\partial t} \tag{D134}$$

which implies that the 3-vector

$$E' = -\nabla V' - \frac{\partial A'}{\partial t}$$

$$= -\nabla V - \nabla \frac{\partial \xi}{\partial t} - \frac{\partial A}{\partial t} + \frac{\partial \nabla \xi}{\partial t} \qquad \text{(D135)}$$

$$= -\nabla V - \frac{\partial A}{\partial t} = E$$

is also locally gauge invariant. Furthermore,

$$\nabla \times \left(E + \frac{\partial B}{\partial t} \right) = -\nabla \times \nabla V = 0 \qquad \text{(D136)}$$

so

$$\nabla \times E = -\frac{\partial (\nabla \times A)}{\partial t} = -\frac{\partial B}{\partial t} \qquad \text{(D137)}$$

which is Faraday's law of induction, another of Maxwell's equations, with E interpreted as the classical electric field. These are, of course, the equations for a free electromagnetic field, that is, one in which the charge and current densities are zero at the point in space where E and B are being calculated.

Summarizing, we have found that the equations that describe the motion of a charged free particle are not invariant under a local gauge transformation. However, we can make them invariant by adding a term to the canonical momentum that corresponds to the 4-vector potential of the electromagnetic field. Thus the electromagnetic force is introduced to preserve local gauge symmetry. Conservation and quantization of charge follow from global gauge symmetry.

THE ELECTROWEAK FORCE

The gauge transformation just described corresponds to a rotation in the abstract space of the 4-momentum eigenstate, which is the state of any particle of constant momentum. Here the transformation operator (D5)

$$U = \exp(i\theta) \qquad \text{(D138)}$$

can be trivially thought of as a 1×1 matrix. The set of all such unitary matrices comprises the transformation group U(1). The generators of the transformation, θ, form a set of 1×1 matrices that, clearly, mutually commute. Whenever the generators of a transformation group commute, that group is termed *abelian*. Electromagnetism is thus an *abelian gauge theory*.

Recall from our discussion of angular momentum that the unitary operator (D76).

$$U = \exp\left(\frac{1}{2} i\boldsymbol{\sigma} \bullet \boldsymbol{\theta} \right) \qquad \text{(D139)}$$

operates in the Hilbert space of spin state vectors. In this case U is represented by a 2×2 matrix. The set of all such matrices comprises the transformation group SU(2), where the prefix S specifies that the matrices of the group are *unimodular*, that is, have unit determinant. This follows from the fact that, for any unitary matrix U,

$$U = \exp(iA) \qquad \text{(D140)}$$

we have

$$\det U = \exp(\mathrm{Tr}A) \qquad \text{(D141)}$$

Since the Pauli matrices are traceless, $\det U = 1$.

Following a procedure similar to what was done above for U(1), let us write

$$U = \exp\left[-\frac{1}{2} i g \boldsymbol{\tau} \times \boldsymbol{\xi} \right] \qquad \text{(D142)}$$

where $\boldsymbol{\theta} = -g\boldsymbol{\xi}$ and where the three components of $\boldsymbol{\tau}$ form a set of matrices identical to the Pauli spin matrices (D77) and we use a different symbol just to avoid confusion with spin. While the spin $\mathbf{S} = \boldsymbol{\sigma}/2$ is a vector in familiar 3-dimensional space, $\boldsymbol{\tau}$ is a 3-vector in some more abstract space we will call *isospin space*. The 3-vector $\mathbf{T} = \boldsymbol{\tau}/2$ is called the *isospin* or *isotopic spin*, analogous to spin angular momentum. It obeys angular momentum commutation rules (D72). and so has the eigenvalue equations,

$$T^2 \left| t, t_3 \right\rangle = t(t+1) \left| t, t_3 \right\rangle$$

(D143)

$$T_3 \left| t, t_3 \right\rangle = t_3 \left| t, t_3 \right\rangle$$

where t is an integer or half-integer and t_3 goes from $-t$ to $+t$ in steps of 1. Thus, t and t_3 are quantum numbers analogous to the spin quantum numbers.

Global gauge invariance under SU(2) implies conservation of isospin. The quarks and leptons of the standard model have $t = 1/2$. The quantity g is a constant analogous to the electric charge, which measures the strength of the interaction.

It is important not to confuse isospin space with the two-dimensional subspace of the state vectors on which U operates. When the isospin space 3-vector $\xi(x)$ depends on the space-time (yet another space) position 4-vector x we again have a local gauge transformation. The generators, being like angular momenta, do not mutually commute, so the group is *nonabelian*. This type of *nonabelian gauge theory* is called a *Yang-Mills theory*.

Let us attempt to make this clearer by rewriting U with indices rather than boldface vector notion:

$$U = \exp\left[-\frac{1}{2} ig \tau^k \xi_k(x) \right]$$

(D144)

where the repeated Latin index k is understood as summed from 1 to 3.

Encouraged by our success in obtaining the electromagnetic force from local U(1) gauge symmetry, let us see what we can get from local SU(2) symmetry. Following the U(1) lead, we define a covariant derivative

$$\mathcal{D}_\mu = \partial_\mu + \frac{1}{2} ig \tau_k W_\mu^k$$

(D145)

where W_μ^k are three 4-vector potentials analogous to the electromagnetic 4-vector potential A_μ. As before, the introduction of the fields W_μ^k maintains local gauge invariance. Or, we can say that local gauge invariance implies the presence of three 4-vector potentials W_μ^k. In the standard model, these are interpreted as the fundamental fields of the weak interaction. Note that the interaction will include charge exchange, as observed experimentally.

In quantum field theory, a particle is associated with every field, the so-called quantum of the field. The spin and parity of the particle, usually written J^P, are determined by the transformation properties of the field. The quantum of a scalar field has $J^P = 0^+$; a vector field has $J^P = 1^-$. For the electromagnetic field described by the potential A_μ, the quantum is the photon. Since A_μ is a vector field, the photon has spin-1. It is a *vector gauge boson.*

Similarly, the weak fields W_μ^k will have three spin-1 particles as their quanta—three vector gauge bosons W$^-$, W^0, and W$^+$; where the superscripts specify the electric charges of the particles. These can also be viewed as the three eigenstates of a particle with isospin quantum number $t = 1$.

If the U(1) symmetry of electromagnetism and the SU(2) symmetry of the weak interaction were perfect, we would see the photon and three W-bosons above. However, these symmetries are broken at the "low" energies where most physical observations are made, including those at the current highest-energy particle accelerators. This symmetry breaking leads to a mixing of the electromagnetic and weak forces. Here, briefly, is how this comes about in what is called *unified electroweak theory.*

The covariant derivative for electroweak theory is written

$$\mathcal{D}_\mu = \partial_\mu + ig_1 \frac{Y}{2} B_\mu + ig_2 \frac{\tau_k}{2} W_\mu^k \qquad \text{(D146)}$$

where the U(1) field is called B_μ and the "coupling constant" g_1 replaces the electric charge in that term. The quantity Y is a constant called the hypercharge generator that can take on different values in different applications, a detail that need not concern us here. The SU(2) term includes a constant g_2; the vector $\mathbf{T} = \boldsymbol{\tau}/2$, or isospin; and the vector field W_μ^k, $k = 1$, 2, 3. We also write these fields W_μ^-, W_μ^0, and W_μ^+ to indicate the electric charges of the corresponding quanta.

Neither the B nor W^0 vector bosons, the quanta of the fields B_μ and W_μ^0, appear in experiments at current accelerator energies. Instead, the particles that do appear are the photon, γ, and Z, whose respective fields A_μ and Z_μ are mixtures of B_μ and W_μ^0. These, together with the W$^+$ and W$^-$, the quanta of the fields W_μ^k, $k = 1$, 2, 3, constitute the vector gauge bosons of the electroweak sector of the standard model. Their mixing is also like a rotation,

$$
\begin{pmatrix} A_\mu \\ Z_\mu \end{pmatrix} = \begin{pmatrix} \cos\theta_w & \sin\theta_w \\ -\sin\theta_w & \cos\theta_w \end{pmatrix} \begin{pmatrix} B_\mu \\ W_\mu^o \end{pmatrix}
$$
(D147)

where the rotation angle θ_w is called the *Weinberg* (or *weak*) *mixing angle*. This parameter is not determined by the standard model and must be found from experiment. The current value (at this writing) is $\sin^2\theta_\omega = 0.23115$. The coupling constants that determine the strengths of the two interactions are

$$
g_1 = \frac{e}{\cos\theta_w} \quad g_2 = \frac{e}{\sin\theta_w}
$$
(D148)

where e is the unit electric charge.

Note that the rotation in the subspace of neutral boson states is a gauge transformation and gauge symmetry would imply that any value of θ_w is possible. The fact that a specific value is evident in the data implies that gauge symmetry is broken in the electroweak interaction.

As we have seen, gauge symmetry requires that the masses of spin-1 bosons be exactly zero. While the photon is massless, the $W^\pm$ and Z bosons have large masses. In the standard model, these masses are understood to arise from a symmetry-breaking process called the *Higgs mechanism*. The symmetry breaking is apparently *spontaneous,* that is, not determined by any known deeper physical principle. Spontaneous symmetry breaking describes a situation, like the ferromagnet, where the fundamental laws are symmetric and obeyed at higher energy, but the lowest-energy state of the system breaks the symmetry. We will discuss spontaneous symmetry breaking and the Higgs mechanism in the following supplement.

THE STRONG FORCE

Moving beyond the weak interactions and SU(2), we have the strong interactions and SU(3). In general, for SU(n) there are $n^2 - 1$ dimensions in the Hilbert subspace. Thus, for $n = 3$, we have eight dimensions. Let us add the new term to the covariant derivative:

$$\mathcal{D}_\mu = \partial_\mu + ig_1 \frac{Y}{2} B_\mu + ig_2 \frac{\tau_k}{2} W_\mu^k + ig_3 \frac{\lambda_a}{2} G_\mu^a \qquad \text{(D149)}$$

where $\mu = 0, 1, 2, 3$ for the four dimensions of space-time, the repeated index k is summed from 1 to 3 in the SU(2) term, and the repeated index a is summed from 1 to 8 in the $SU(3)$ term. The λ_a are eight traceless 3×3 matrices analogous to the three Pauli 2×2 isospin matrices τ_k, and the G_μ^a are eight spin-1 fields analogous to the singlet field B_μ and the triplet field W_μ^k of the electroweak interaction. The gauge bosons in this case are eight *gluons*. The symmetry is not broken, so the gluons are massless. Global gauge invariance under SU(3) implies the conservation of another quantity called *color charge*.

NOTE

1. See, for example, J. J. Sakurai, *Modern Quantum Physics* (New York: Addison-Wesley, 1985), pp. 34–36.

MATHEMATICAL SUPPLEMENT E

THE STANDARD MODEL

SUMMARY

The basic idea of the standard model can be summarized as follows. Whenever an elementary physics expression, such as a Lagrangian density or an equation of motion, contains a partial derivative operator $\partial_\mu = \partial/\partial x_\mu$ we simply replace it with the covariant derivative

$$\mathcal{D}_\mu = \partial_\mu + ig_1 \frac{Y}{2} B_\mu + ig_2 \frac{\tau_k}{2} W_\mu^k + ig_3 \frac{\lambda_a}{2} G_\mu^a \qquad (E1)$$

where $g_1 = e/\cos\theta_w$, $g_2 = e/\sin\theta_w$, e is the unit electric charge, θ_w is the Weinberg or weak mixing angle (about 28° from experiment), g_3 is the strong interaction strength, τ_k are three traceless 2×2 matrices identical to the Pauli spin matrices, λ_a are eight traceless 3×3 matrices, and Y is a numeric constant, the *hypercharge*,

$$Y = 2q - t_3 \qquad (E2)$$

where q is the charge and t_3 is the third component of the isospin of the particle.

The weak bosons $W^\pm$ have charges $\pm e$ and comprise the quanta of the fields $W_\mu^\pm$. The photon, symbol γ, is the quantum of the field A_μ and the Z-boson is the quantum of the field Z_μ, each with zero charge. The neutral fields mix according to

$$\begin{pmatrix} A_\mu \\ Z_\mu \end{pmatrix} = \begin{pmatrix} \cos\theta_w & \sin\theta_w \\ -\sin\theta_w & \cos\theta_w \end{pmatrix} \begin{pmatrix} B_\mu \\ W_\mu^o \end{pmatrix} \qquad (E3)$$

That is, B_μ and W_μ^o are the fundamental fields while A_μ and Z_μ appear as the observable fields at the low energy of current experimentation (and common experience).

The quantities e, θ_w, and g_3 are all determined by experiment. The photon and gluon are required to have zero mass in the model, but the weak bosons have nonzero masses.

QUARKS AND LEPTONS

First, consider the U(1) symmetry of the electromagnetic interaction. This symmetry implies that each quark and lepton carries an electric charge (or zero charge) that will be conserved. The values of those charges are not determined by the model and must be obtained by experiment. Second, consider the SU(2) symmetry of the weak interaction. This implies that the quarks and leptons will each have a conserved isospin $t = 1/2$ and so will appear in doublets with $t_3 = \pm 1/2$.

That is, they will have isospin eigenstates $\left| t, t_3 \right\rangle = \left| \dfrac{1}{2}, \pm\dfrac{1}{2} \right\rangle$.

Third, consider the SU(3) symmetry of the strong interaction. This implies that each quark (leptons do not participate in the strong interaction) carries another conserved quantity that can have three possible values. That quantity is called *color*, or *color charge*, in analogy with the familiar color phenomenon of everyday life. Thus, the quarks are "red," "green," or "blue" (r, g, b), analogous to the primary colors. All three add up to white, that is, neutral color charge. The antiquarks have the "anti-colors $\bar{r}, \bar{g}, \bar{b}$," which we can think of as cyan, magenta, and yellow—the colors you get when you filter red, green, or blue light, respectively, from a beam of white light. The anticolor vectors point in the opposite direction to the color vectors in color space. And so $r\bar{r}$, $g\bar{g}$, and $b\bar{b}$ all give white. Other quark-antiquark combinations can be formed, such as $\bar{r}b$, that are not colorless, that is, not white. The eight gluons of the strong force carry these colors, where the basic interaction is $\bar{q} + q \leftrightarrow G$. While these may be exchanged as virtual particles in fundamental reactions, stable states of color do not seem to occur in nature.

In another experimental fact not directly predicted by the standard model, quark and lepton doublets appear in three *generations*. The first generation quarks are the *up* quark, isospin state

$$u = \left| \frac{1}{2}, \frac{1}{2} \right\rangle, \text{ and } down \text{ quark, } d = \left| \frac{1}{2}, -\frac{1}{2} \right\rangle$$

("up" and "down" in weak isospin space). These comprise the main ingredients of the proton and neutron: the $p = uud$, the $n = udd$. The first-generation leptons are the *electron neutrino*,

$$\nu_e = \left| \frac{1}{2}, \frac{1}{2} \right\rangle \text{ and } electron, \ e^- = \left| \frac{1}{2}, -\frac{1}{2} \right\rangle.$$

Quarks and leptons are distinguished from one another by the fact that quarks participate in strong interactions while leptons do not. Strongly interacting particles are called *hadrons*. Both quarks and leptons participate in the electromagnetic and weak, or *electroweak*, interactions.

The three generations of quarks and leptons can be represented as follows:

Quarks   **Leptons**

$$\begin{pmatrix} u \\ d \end{pmatrix} \begin{pmatrix} c \\ s \end{pmatrix} \begin{pmatrix} t \\ b \end{pmatrix} \qquad \begin{pmatrix} \nu_e \\ e^- \end{pmatrix} \begin{pmatrix} \nu_\mu \\ \mu^- \end{pmatrix} \begin{pmatrix} \nu_\tau \\ \tau^- \end{pmatrix}$$

and their antiparticles. The quarks on the top line in each generation have $t_3 = +1/2$ and electric charge $+2e/3$, where e is the unit electric charge. The quarks on the bottom line have $t_3 = -1/2$ and charge $-e/3$. The second generation quarks are called, for historical reasons, *charmed* and *strange*. The third-generation quarks are called *top* and *bottom* or, sometimes, "truth" and "beauty."

The leptons on the top line in each generation have $t_3 = +1/2$ and electric charge zero. The leptons on the bottom line have $t_3 = -1/2$ and electric charge $-e$. The second- and third-generation charged leptons are the *muon* and *tauon*, which are little more than heavier versions of the electron. Each is accompanied in its weak isospin doublet by a neutrino—the *mu* and *tau* neutrinos.

Most of the visible matter of the Universe is composed of just the three particles: u, d, and e. However, many short-lived particles are formed from the other quarks and leptons. A hundred or more unstable hadrons heavier than protons and neutrons are now known to be formed from three quarks selected from any of the three generations. These are half-integer-spin fermions called *baryons.* Another hundred or so hadrons are formed from quark-antiquark pairs. They are integer- or zero-spin bosons called *mesons.* Although the quarks are fractionally charged, no composite particles of fractional charge have ever been observed. Similarly, although the quarks have different color charges, only white composite particles have been observed. Color helps maintain the Pauli exclusion principle for particles, like the proton and neutron, which have two or more otherwise identical fermions in the same state. Each of the three quarks in a baryon is in a different color state, adding up to white.

Each succeeding generation of quarks and leptons is more massive than the next. The neutrinos, once thought to be massless, have now been found to have very tiny masses. There are reasons to believe from both cosmology and accelerator experiments that no more than three generations exist.

THE RANGES OF FORCES

To see the dependence the range of a force has on the mass of the exchanged particle, let us use the argument made by Hideki Yukawa back in 1934 when he introduced the meson model of nuclear forces. In classical electrodynamics, the scalar potential for a given time-independent charge distribution $q\rho(\mathbf{r})$ is obtained as the solution of Poisson's equation (Gaussian units used here for simplicity)

$$\nabla^2 V = -4\pi q \rho(\mathbf{r}) \tag{E4}$$

This has the general solution

$$V(\mathbf{r}) = \int d^3\mathbf{r}' \frac{q\rho(\mathbf{r}')}{|\mathbf{r} - \mathbf{r}'|} \tag{E5}$$

which gives for a point source, $\rho(\mathbf{r}') = \delta(0)$,

$$V(\mathbf{r}) = \frac{q}{r} \tag{E6}$$

which is the familiar Coulomb potential. Yukawa modified this, proposing that the potential Φ of the nuclear force is a solution of

$$\left(\nabla^2 - k^2\right)\Phi = 4\pi g\rho(\mathbf{r}) \tag{E7}$$

where g is a constant that determines the strength of the nuclear force, analogous to the electric charge. The general solution is

$$\Phi(\mathbf{r}) = -g\int \frac{\exp\left(-k|\mathbf{r} - \mathbf{r}'|\right)}{|\mathbf{r} - \mathbf{r}'|}\rho(\mathbf{r}')d^3\mathbf{r}' \tag{E8}$$

which, for a nuclear point source $\rho(\mathbf{r}') = \delta(0)$, gives

$$\Phi(\mathbf{r}) = -g\frac{\exp\left(-kr\right)}{r} \tag{E9}$$

Note that the wave function of a free spin-0 particle in relativistic quantum mechanics is given by the Klein-Gordon equation

$$H^2\Phi = \left(P^2 + m^2\right)\Phi \tag{E10}$$

or, after putting in the explicit differential operators for H and P,

$$-\frac{\partial^2\Phi}{\partial t^2} = \left(-\nabla^2 + m^2\right)\Phi \tag{E11}$$

which gives for the time-independent function $\Phi(\mathbf{r})$

$$\left(\nabla^2 - m^2\right)\Phi(\mathbf{r}) = 0 \tag{E12}$$

This shows that $k = m$, the mass of a spinless particle that is the quantum of the nuclear potential field Φ, and

$$\Phi(\mathbf{r}) = -g\frac{\exp\left(-mr\right)}{r} \tag{E13}$$

We see that this potential, called the *Yukawa potential*, falls off faster than the Coulomb potential that we associate with the massless photon. In units, $\hbar = c = 1$, the range of the nuclear force is given by $R = 1/m$. In more familiar units,

$$R = \frac{\hbar}{mc} = \frac{197}{mc^2}\,\text{fm} \tag{E14}$$

where mc^2 is in MeV. Yukawa estimated the range of the nuclear force to be 1.4 fm, thus predicting the existence of a particle with mass $m = 140$ MeV/c^2. A particle of this mass, the pion, was discovered in 1947.

SPONTANEOUS SYMMETRY BREAKING

Let us look at a simple model for spontaneous symmetry breaking and see how that leads to a field whose quantum has mass and how this quantum, the *Higgs boson*, provides for the mechanism by which the weak bosons gain mass in the standard model. Consider a scalar field ϕ. The Lagrangian density of the field is

$$\mathcal{L} = \frac{1}{2}\partial_\mu\phi\partial^\mu\phi - u(\phi) \tag{E15}$$

Suppose that the potential energy has the form of a power law

$$u(\phi) = \frac{1}{2}\mu^2\phi^2 + \frac{1}{4}\lambda\phi^4 \tag{E16}$$

where we assume $u(-\phi) = u(\phi)$. This is illustrated in fig. E5.1. When $\mu^2 > 0$ we have a single minimum at $\phi = 0$. When $\mu^2 < 0$ we have a maximum at $\phi = 0$ and a minimum at $\phi = \pm v$ where

$$v = \pm\left(-\frac{\mu^2}{\lambda}\right)^{1/2} \tag{E17}$$

We can picture a phase transition taking us from an initially symmetric state with zero field $\phi = 0$, to one with broken symmetry and nonzero field $\phi = \pm v$.

Note that the potential energy at this minimum is negative

$$u(v) = -\frac{\mu^4}{4\lambda} \tag{E18}$$

This is the lowest-energy state and hence, the vacuum state. The field is nonzero in the vacuum and its value v at this point is called the *vacuum*

$$u(\phi) = \mu^2\phi^2/2 + \lambda\phi^4/4$$

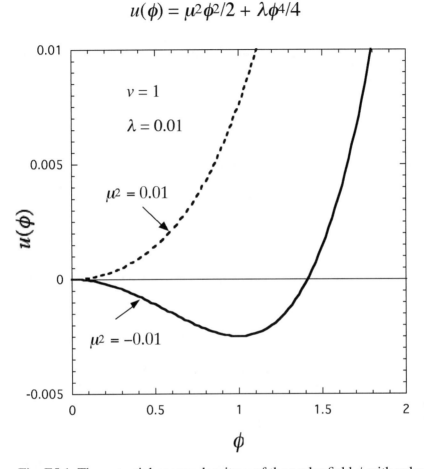

Fig. E5.1. The potential energy density u of the scalar field ϕ with values of parameters chosen for illustration, for the two cases where μ^2 is positive or negative. When negative, the minimum is at a nonzero value of the field. The vacuum expectation value, where u is minimum, is $v = 1$ in this specific example.

expectation value. Since we can always add an arbitrary constant to the potential energy, we can also write

$$u(\phi) = \frac{1}{2}\mu^2\phi^2 + \frac{1}{4}\lambda\phi^4 + \frac{\mu^4}{4\lambda} \tag{E19}$$

so that $u(v) = 0$. However, let us not bother to do that at this point. Then we have

$$\mathcal{L} = \frac{1}{2}\partial_\mu \phi \partial^\mu \phi - \frac{1}{2}\mu^2\phi^2 - \frac{1}{4}\lambda\phi^4 \tag{E20}$$

Let us write

$$\phi(x) = v + h(x) \tag{E21}$$

where $h(x)$ is the displacement of $\phi(x)$ away from the vacuum state. Then, in the vicinity of the minimum (dropping terms in h^3 and higher; the first-order terms in h cancel), some algebra will show that

$$\mathcal{L} = \frac{1}{2}\partial_\mu h\partial^\mu h - \lambda v^2 h^2 + \text{constant} \tag{E22}$$

This is just the Klein-Gordon Lagrangian density for a spinless particle of mass m_h, where

$$m_h^2 = 2\lambda v^2 = -2\mu^2 \tag{E23}$$

This illustrates the Higgs mechanism by which masses are generated by spontaneous symmetry breaking. The vacuum contains a scalar field h, which can be thought of as composed of spin-0 particles of mass m_h, whose value is undetermined in the model. These Higgs bosons can be thought of as filling all of space.

Now the above Lagrangian density is not locally gauge invariant. It can be made locally gauge invariant by replacing ∂_μ by

$$\mathcal{D}_\mu = \partial_\mu + igA_\mu \tag{E24}$$

where A_μ is a vector field—a gauge field that is then introduced in the process. The quanta of this field are particles of spin-1. The Lagrangian density for massless spin 1 particles is, from supplement D,

$$\mathcal{L} = -\frac{1}{16\pi}F^{\mu\nu}F_{\mu\nu} \tag{E25}$$

where

$$F_{\mu\nu} = \partial_\mu A_\nu - \partial_\nu A_\mu \tag{E26}$$

and is also gauge invariant.

Suppose, then, that we have a mixture of spin-0 Higgs bosons and spin-1 gauge bosons. The Lagrangian density will be, in the region near the potential minimum,

$$\mathcal{L} = \frac{1}{2}\left[\left(\partial_\mu - igA_\mu\right)\phi\right]\left[\left(\partial^\mu + igA^\mu\right)\phi\right]$$
$$-\frac{1}{2}\mu^2\phi^2 - \frac{1}{4}\lambda\phi^4 - \frac{1}{16\pi}F_{\mu\nu}F^{\mu\nu} \tag{E27}$$

Again, let us write $\phi(x) = v + h(x)$ and drop higher-order terms in h. The first-order terms again cancel. Note that in the standard model, ϕ is taken to be complex; here I am making a simplification to make the essential point and not attempting to reproduce the exact standard model. Then,

$$\mathcal{L} = \frac{1}{2}\partial_\mu h\partial^\mu h + \frac{1}{2}g^2 hA_\mu A^\mu - \frac{1}{2}m_h^2 h^2$$
$$+\frac{1}{2}g^2 v^2 A_\mu A^\mu - \frac{1}{16\pi}F_{\mu\nu}F^{\mu\nu} \tag{E28}$$

Recalling the Proca-Lagrangian density for a spin-1 particle of mass m_A,

$$\mathcal{L} = -\frac{1}{16\pi}F_{\mu\nu}F^{\mu\nu} + \frac{1}{8\pi}m_A^2 A_\mu A^\mu \tag{E29}$$

we see that the our new Lagrangian describes a spin-0 particle of mass m_h and a spin-1 particle of mass

$$m_A = 2\sqrt{\pi}gv \tag{E30}$$

In the full standard model, which includes electroweak unification, we get three massive vector gauge bosons along with a massless photon (to preserve U(1) gauge symmetry) and a massive spin-0 Higgs boson.[1]

NOTE

1. For a good exposition that does not use any more high-level mathematics than assumed here, see the last chapter of F. Mandl and G. Shaw, *Quantum Field Theory*, rev. ed. (New York: Wiley, 1984).

MATHEMATICAL SUPPLEMENT F

STATISTICAL PHYSICS

THERMODYNAMICS

The smallness of atoms means that huge numbers of them are present in even the tiniest amount of matter. Thus the statistical fluctuations are too small to be observed in everyday phenomena. For example, consider the average pressure p on the walls of a chamber of volume V containing N atoms at an absolute temperature T given by the well-known ideal gas equation of state, which we will derive later

$$p = \frac{NkT}{V} \tag{F1}$$

where k is Boltzmann's constant. Let $V = 1$ m^3 and $T = 300$ K. Then $p = N(1.38 \times 10^{-23})(300) = (4.14 \times 10^{-21})N$ in Newtons/m^2. The statistical fluctuation in this quantity will be characterized by a standard deviation $\Delta p = (4.14 \times 10^{-21})(N + 1)^{1/2}$. If, for example, $N = 100$, the pressure will be $p = 4.14 \times 10^{-19} \pm 4.16 \times 10^{-20}$. That is, the pressure fluctuates by 10 percent about its mean. For typical macroscopic systems, however, N is on the order of 10^{24}. In that case, $p = 4.14 \times 10^3 \pm 4.14 \times 10^{-9}$. The fluctuation of one part in a trillion in this case is not likely to be noticed. The practical result is that, for most purposes, thermodynamic quantities such as the average pressure can be treated as if they are exact.

The first law of thermodynamics describes the special form conservation of energy takes when dealing with heat phenomena. The heat Q added to a system is equal to the change ΔU in internal energy of the system, plus the work W done by the system.

273

$$Q = \Delta U + W \tag{F2}$$

Two systems in contact can exchange energy back and forth in the form of heat or work.

Note the convention that W is positive when the work is done by the system. It will be negative when work is done on the system. In general,

$$W = \int_{o}^{\Delta V} p\,dV \tag{F3}$$

where ΔV is the change in the volume V of the system. Normally p is positive (but not always), so work is done by a system, such as a gas, when it expands.

The internal energy of a body is proportional to its absolute temperature and represents stored energy that can be used to do work or transmit heat to another body. For example, if $Q = 0$, then $W = -\Delta U$ and work done by the system will be accompanied by a decrease in internal energy. In another example, $W = 0$ and $Q = \Delta U$. In this case, the heat added to a body goes into increasing its internal energy. If $Q < 0$, heat is removed from the body.

The first law does not forbid the energy in our surrounding environment from being used to run a heat engine. But our experience tells us that we have to regularly pull into a filling station to fuel our cars; perpetual motion machines do not exist.

The second law of thermodynamics was introduced in the nineteenth century in order to describe these empirical facts. Sadi Carnot (d. 1832) provided a mathematically precise form of the second law by stating it in terms of a quantity called the entropy. The change in entropy of a system as it goes from one thermodynamic state to another is defined as

$$\Delta S = \int_{0}^{Q} \frac{dQ}{T} \tag{F4}$$

where T is the absolute temperature and the integral is over any path that takes you from one state to another. The thermodynamic state of a system is defined by a complete set of variables needed to define the system, such as the pressure, temperature, and density of a gas.

The fact that the entropy can be calculated along any path is an important one. It means that you can calculate ΔS for a nonequilibrium path, where T is undefined, by using any equilibrium or quasi-equilibrium path that connects the initial and final states. For example, for any process in which the initial and final temperatures are the same, no matter what they may be in between, we can calculate the change in entropy using an isothermal (constant T) process:

$$\Delta S = \frac{Q}{T} \tag{F5}$$

where $Q = \int dQ$ is the heat added (subtracted if negative) during the isothermal process.

In terms of entropy, the second law then says that the entropy of an isolated system must remain constant or increase with time. It can never decrease.

Suppose an infinitesimal amount of heat dQ is exchanged between two bodies of absolute temperatures T_1 and T_2. Let dQ be positive and assume the heat flows from body 2 to body 1. The two bodies together form an isolated system. In that case,

$$dS = \frac{dQ}{T_1} - \frac{dQ}{T_2} = dQ\left(\frac{1}{T_1} - \frac{1}{T_2}\right) \tag{F6}$$

and $dS > 0$ requires that $T_2 > T_1$.

STATISTICAL MECHANICS

The pressure on the walls of a chamber containing a fluid was seen to be the average effect of the molecules of the fluid hitting the walls and transferring momentum to those walls. Assuming a billiard-ball model for the atoms of an ideal gas, the familiar equation of state

$$p = \frac{NkT}{V} \tag{F7}$$

is easily derived (see below). The internal energy of a monatomic ideal gas is also shown to be simply

$$U = N\left\langle\frac{1}{2}mv^2\right\rangle = \frac{3}{2}NkT \tag{F8}$$

where $\left\langle \dfrac{1}{2}mv^2 \right\rangle$ is the average kinetic energy of a molecule in the gas.

This identifies the absolute temperature T with the average kinetic energy.

The second law of thermodynamics follows from *Boltzmann's H-theorem*, which is proved in advanced textbooks on statistical mechanics, such as the classic monograph *The Principles of Statistical Mechanics* by Richard Tolman.[1] Here we will just outline the idea.

Let us begin by noting that in classical mechanics the state of a system of N particles can be specified as a point in an abstract, multidimensional space called *phase space*. In the simplest case in which the particles are identical and have no rotational, vibrational, and other internal degrees of freedom, that phase space has $6N$ dimensions, $3N$ dimensions to describe the position of each particle and $3N$ dimensions to describe their momenta. That is, each particle has three degrees of freedom and each of these is represented by the motion of a point in a two-dimension phase space, where one coordinate is x and the other is its conjugate momentum p_x. This can be generalized to include other degrees of freedom, which can also be described in terms of coordinates and their conjugate momenta.

Boltzmann defined a quantity H by

$$H = \sum_{i=1}^{D} P_i \ln P_i = \left\langle \ln P_i \right\rangle \tag{F9}$$

where P_i is the probability for finding a system of particles in a particular state i, that is, at a particular point in phase space, and D is the total number of degrees of freedom or dimension of phase space. He then derived a theorem, Boltzmann's H-theorem, which shows that a molecular system originating arbitrarily far away from equilibrium will approach equilibrium, and that H will be minimum at that point.

Let us define a quantity

$$S = -kH = -k \left\langle \ln P_i \right\rangle \tag{F10}$$

which reaches a maximum at equilibrium. In the case in which a system has a total of Ω states, each of equal probability, the probability for a given state is then simply $1/\Omega$ and the entropy (F10) becomes

$$S = -k \ln\left(\frac{1}{\Omega}\right) = k \ln\Omega \qquad \text{(F11)}$$

In a moment I will show that (F10) is identical to the thermodynamic entropy. Given this is the case, Boltzmann's *H*-theorem amounts to a derivation of the second law of thermodynamics. A system of interacting particles will eventually reach maximum entropy.

Suppose the particles of a system move around randomly, colliding with one another, but are generally unconstrained except by conservation of energy, momentum, and other conservation laws. We can picture these laws as forming a boundary in phase space so that all the allowed states are within that boundary. If no other dynamical principles are in action, then all the states in the allowed phase space are equally likely.

The states of the system are points within the allowed phase space. Let the total number of states be Ω. Classically, this is infinite. However, quantum mechanics implies that each two-dimensional subspace of the phase space volume is divided into cells of area, $\Delta p \Delta x = h$, where h is Planck's constant, for each degree of freedom.[2]

If we assume no internal degrees of freedom such as spin, Ω is then equal to the volume of allowed phase space divided the cell size. That is,

$$\left(\Delta p \Delta x\right)^{3N} = h^{3N} \qquad \text{(F12)}$$

so

$$\Omega = \left(\text{volume of allowed phase space}\right) / h^{3N} \qquad \text{(F13)}$$

This can be multiplied by the number of internal degrees of freedom. For example, if the particle has spin quantum number s, multiply by $2s + 1$.

Suppose we have a single, spinless nonrelativistic particle of mass m with total energy E within a volume V.

$$E = \sum_{i=1}^{3} \frac{p_i^2}{2m} \qquad \text{(F14)}$$

where p_i is the momentum for the ith degree of freedom and i goes from 1 to 3. Consider first all the states with energy less than E. In that case, the allowed phase space for this system will be composed of a three-dimensional configuration subspace volume V and a three-dimensional spherical momentum subspace with a radius $(2mE)^{1/2}$. The total number of accessible states will be

$$\Omega(< E) \sim V E^{3/2} \qquad \text{(F15)}$$

If we have N particles, this becomes

$$\Omega(< E) \sim V^N E^{3N/2} \qquad \text{(F16)}$$

Finally, the number of accessible states of energy E per unit energy will be

$$\Omega(E) = \frac{\partial \Omega(< E)}{\partial E} \sim V^N E^{3N/2-1} \qquad \text{(F17)}$$

In this case, we see that

$$S = k \ln \Omega = N \ln V + \ln \xi(E) \qquad \text{(F18)}$$

where $\xi(E)$ is a function of E. This general form will hold whenever the volume of phase space can be factorized into a configuration subspace and momentum subspace. Since N is generally a large number, the second term in (F18) is negligible and the entropy is proportional to the number of particles.

This result will also hold in another general class of applications. Consider a system of N identical particles. Let n be the number of accessible states for one particle. The total number of accessible states is then $\Omega = nN$ and the entropy

$$S = k \ln\left(n^N\right) = kN \ln(n) \qquad \text{(F19)}$$

For example, consider three first-generation quarks, each of a different "color" so we need not worry about the Pauli exclusion principle. Each quark has two weak isospin states, "up" or "down." The possible color states of the system are *uuu*, *uud*, *udu*, *udd*, *duu*, *dud*, *ddu*, and *ddd*. That is, $n^N = 2^3 = 8$ and $S = 3k\ln2$.

We can now show how statistical mechanics leads to the principles of classical thermodynamics discussed earlier in this supplement. Consider two systems A and A' with energies E and E' in contact with one another but otherwise isolated. From energy conservation, $E' = E_0 - E$, where E_0 is the constant total energy. The probability for an energy E will be

$$P(E) = C\Omega(E)\Omega'(E_0 - E) \tag{F20}$$

where $\Omega(E)$ is the number of states accessible to A with energy E, $\Omega'(E_0 - E)$ is the number of states accessible to A' with energy $E_0 - E$, and C is a constant. We can write

$$\ln P(E) = \ln C + \ln \Omega(E) + \ln \Omega'(E_0 - E) \tag{F21}$$

This will be maximum when

$$\frac{\partial \ln P(E)}{\partial E} = \frac{\partial \ln \Omega(E)}{\partial E} + \frac{\partial \ln \Omega'(E_0 - E)}{\partial E} = 0 \tag{F22}$$

or when

$$\frac{\partial S}{\partial E} - \frac{\partial S'}{\partial E'} = 0 \tag{F23}$$

That is, maximum probability is equivalent to maximum entropy, or minimum H as required by Boltzmann's H-theorem.

We define the absolute temperature T by

$$\frac{1}{T} = \frac{\partial S}{\partial E} = \frac{\partial S'}{\partial E'} \tag{F24}$$

Identifying this maximum probability situation with thermal equilibrium, we thus prove the zeroth law of thermodynamics in which two systems in thermal equilibrium have the same temperature.

The first law follows from energy conservation

$$dE = dQ - dW \tag{F25}$$

Let us consider the case where the only work results from volume changes, that is, $dW = pdV$ (see the textbooks for the more general case). Then

$$dE = dQ - p\Delta V \tag{F26}$$

We can take $\Omega = \Omega(E, V)$, so that

$$d\ln\Omega = \frac{\partial\ln\Omega}{\partial E}dE + \frac{\partial\ln\Omega}{\partial V}dV \tag{F27}$$

which we can write

$$dS = \frac{\partial S}{\partial E}dE + \frac{\partial S}{\partial V}dV \tag{F28}$$

and thus, since $\dfrac{1}{T} = \dfrac{\partial S}{\partial E}$,

$$dE = TdS - T\frac{\partial S}{\partial V}dV \tag{F29}$$

from which we make the connections

$$dQ = TdS \tag{F30}$$

and

$$p = T\frac{\partial S}{\partial V} \tag{F31}$$

The first, (F30), is the familiar definition of entropy in classical thermo-dynamics. The second, (F31), is an expression that allows us to calculate the equation of state of a system. In the earlier example of N nonrela-tivistic particles in equilibrium in chamber of volume V with a total energy E, we found that that the entropy is given by (F18),

$$S = k\ln\Omega = N\ln V + \ln\xi(E) \tag{F32}$$

so, from (F31),

$$p = \frac{NkT}{V} \tag{F33}$$

the ideal gas equation of state.

ENTROPY AND INFORMATION

An important connection exists between entropy and information in the practice of modern computer and information science.

Suppose we want to transmit a message containing a single symbol, such as a letter or a number, from a set of n symbols. In 1948 Claude Shannon defined a quantity

$$H_S = -\sum_{i=1}^{n} P_i \log_2 P_i = -\langle \log_2 P_i \rangle \qquad \text{(F34)}$$

which he called "the entropy of the set of probabilities $P_1 \ldots P_n$" for the symbols in the message.[3] That is, P_i is the probability for the ith symbol in the list. Because of the base-2 logarithm, the units of H_S are bits.

In today's literature on information theory, H_S is called the *Shannon uncertainty*. The information I carried by a message is defined as the decrease in Shannon uncertainty when the message is transmitted. That is,

$$I = H(\text{before})\text{-}H(\text{after}) \qquad \text{(F35)}$$

If we consider the special case when all the probabilities P_i are equal to P, we get the simpler form

$$H_S = -\log_2 P \qquad \text{(F36)}$$

Let me illustrate the idea of information with a simple example of a single-character message that can be one of the eight letters S, T, U, V, W, X, Y, or Z with equal probability. Before the message is transmitted, $n = 8$, $P = 1/8$, and $H(\text{before}) = -\log_2(1/8) = \log_2(8) = 3$. After the message is successfully transmitted, we know what the character is, so $P = 1$ and $H(\text{after}) = -\log_2(1) = 0$. Thus $I = 3$ bits of information that are received as the uncertainty is reduced by three bits.

Now suppose that the message is a little garbled so that we know the symbol transmitted is either a U or a V, but we cannot tell which, and they have equal probability. Then, after the message is received, $P = 1/2$ and $H(\text{after}) = -\log_2(1/2) = 1$. In that case, $I = 3 - 1 = 2$ bits of information are received.

Shannon noted that "the form of H [what I call H_S] will be recognized as that of entropy as defined in certain formulations of statistical mechanics," which was discussed above. Shannon explicitly states that "H is then, for example, the H in Boltzmann's H-theorem." Actually, as we saw above, in statistical mechanics the quantity H is conventionally defined without the minus sign and uses the natural logarithm

$$H = \sum_i P_i \ln P_i \tag{F37}$$

However, as Shannon noted, any constant multiplying factor, positive or negative, could have suited his purposes since that just sets the units. As we saw above, the choice he made gives units conveniently in dimensionless bits.

In particular, the quantity H was seen to be simply related to the thermodynamic entropy S by $S = -kH$, where k is Boltzmann's constant. The relationship between the entropy S of statistical mechanics and the Shannon uncertainty H_S then is,

$$S = k \ln(2) H_S \tag{F38}$$

That is, they are equal within a constant and they have the same sign. So, Shannon was justified in calling H the *entropy*.

For the rest of this book I will adopt Shannon's more sensible definition of entropy, that is, let $k\ln(2) = 1$, so that the units of entropy are dimensionless bits.

$$S = H_S = -\sum_{i=1}^{n} P_i \log_2 P_i = -\left\langle \log_2 P_i \right\rangle \tag{F39}$$

When all the P_is are equal to $1/\Omega$, where Ω is the total number of accessible states of the system,

$$S = \log_2 \Omega \tag{F40}$$

In this way we can directly identify changes in the entropy of a system with changes in the system's information content. We can also associate entropy with the uncertainty of our knowledge of the structure of a system giving precise, mathematical meaning for entropy as a measure of "disorder."

RADIOACTIVE DECAY

Let $\lambda(t)$ be the probability that a nucleus decays at a time t per unit time interval. Let $N(t)$ be the number of nuclei that existed prior to t. Then the number of nuclei that decay in the time interval dt is

$$dN = \lambda(t)N(t)dt \tag{F41}$$

and

$$\int_{N(0)}^{N(t)} \frac{dN}{N(t)} = \int_0^t \lambda(t)dt \tag{F42}$$

Now, suppose that the decay probability is invariant under time translation. In that case, $\lambda(t) = \lambda = $ a constant and we can easily solve the above integral to get

$$N(t) = N(0)\exp(-\lambda t) \tag{F43}$$

the familiar exponential decay "law." However, the decay is *random* since all time intervals are equally likely. Any observed deviation from exponential decay would be evidence for a nonrandom process involved in nuclear decay.

NOTES

1. Richard C. Tolman, *The Principles of Statistical Mechanics* (London: Lowe & Brydone, 1938).
2. Note that Δp and Δx here are not the standard deviations, but the cell widths. So this is not a direct application of the uncertainty principle, although it is related. Here, simply think of a single wavelength $\lambda = \Delta x = h/p$ in each cell.
3. Claude Shannon and Warren Weaver, *The Mathematical Theory of Communication* (Urbana: University of Illinois Press, 1949).

MATHEMATICAL SUPPLEMENT G

COSMOLOGY

THE EXPANDING UNIVERSE

Hubble found that, on average, the speed v at which a galaxy recedes from us is proportional to its distance r, as one would expect from an explosion in which the faster-moving fragments go farther than slower ones, as given by

$$v = Hr \tag{G1}$$

The proportionality factor H is called the *Hubble constant*; its most recent estimated value is 71 kilometers per second per million parsecs, where a parsec equals 3.26 light-years or 3.06×10^{16} meters.

GENERAL RELATIVITY AND COSMOLOGY

In general relativistic cosmology, the expansion of an assumed homogeneous and isotropic universe is described by a scale factor $a(t)$, which characterizes the distances between bodies as a function of time. The invariant space-time interval is given by the Friedmann-Robertson-Walker metric

$$ds^2 = dt^2 - a(t)^2 \left[\frac{dr^2}{1 - kr^2} + r^2 \left(d\theta^2 + \sin^2 \theta d\phi^2 \right) \right] \tag{G2}$$

where spherical spatial coordinates are used and we work in units where $c = 1$. The universe is spatially flat (Euclidean geometry) if $k = 0$, the curvature of space is positive if $k = +1$, and the curvature is negative if

$k = -1$. The universe is open for $k = 0, -1$, and closed for $k = +1$. Note that, as defined, r is dimensionless and a has the dimensions of length.

Textbooks on general relativity prove that for a universe of average mass-energy density ρ,

$$\left(\frac{da}{dt}\right)^2 - \frac{8\pi G\rho}{3}a^2 = -k \qquad \text{(G3)}$$

and

$$\frac{d^2a}{dt^2} = -\frac{4\pi G}{3}\left(\rho + 3p\right)a(t) \qquad \text{(G4)}$$

where p is the pressure. These are called the Friedmann equations.

In a matter-dominated universe, $p \ll \rho$. To see this, assume the material particles obey the ideal gas equation of state, derived in supplement F,

$$pV = Nk_BT \qquad \text{(G5)}$$

where N is the number of particles in the volume V, k_B is Boltzmann's constant, and T is the absolute temperature. We can write this

$$p = \frac{\rho k_B T}{m} \qquad \text{(G6)}$$

where ρ is the mass density and m is the mass of the particle. The matter-dominated universe is defined as one consisting mainly of nonrelativistic particles. Since $k_B T$ is on the order of the kinetic energy of the particle, $k_B T \ll mc^2$ when $v \ll c$, and so $p \ll \rho c^2 \approx 0$. For matter domination we get the Newtonian form (with $c = 1$ again)

$$\frac{d^2a}{dt^2} = -\frac{4\pi G}{3}\rho a(t) = -\frac{GM}{a^2} \qquad \text{(G7)}$$

where M is the mass enclosed in a sphere of radius a.

In a radiation-dominated universe, defined as one consisting mainly of relativistic particles such as photons, $v \approx c$, the equation of state is

$$p = \frac{\rho}{3} \qquad \text{(G8)}$$

which gives

$$\frac{d^2a}{dt^2} = -\frac{8\pi G}{3}\rho a(t) = -2\frac{GM}{a^2} \qquad (G9)$$

This shows that photons have twice the acceleration of gravity as nonrelativistic matter. In general, ρ and p will include matter and radiation, and whatever else may be contained in a universe that has mass-energy.

ENERGY CONSERVATION

Note that the first of Friedmann's equations says that the quantity

$$\left(\frac{da}{dt}\right)^2 - \frac{8\pi G\rho}{3}a^2 = -k = \text{constant} \qquad (G10)$$

This can be written

$$\tfrac{1}{2}mv^2 - \frac{GmM}{a} = -\tfrac{1}{2}mk \qquad (G11)$$

where $v = da/dt$, which indicates that the kinetic energy plus the gravitational potential energy of a particle of mass m moving with the expansion on a sphere of radius a enclosing a mass M is constant. That is, the total mechanical energy is conserved when the particle is moving at the speed at which the sphere expands, that is, has negligible local motion. In a flat universe, $k = 0$, the total mechanical energy is zero.

However, it should be noted that the quantity $\frac{1}{2}mv^2$ is the kinetic energy of a particle only when $v \ll c$; so it would be wrong to conclude that mechanical energy is conserved in general. The relationship above, however, is still valid for all particle speeds; it contains a conserved quantity that is not in general equivalent to the mechanical energy.

It should not be surprising that mechanical energy is not conserved at relativistic speeds. This is indeed the case for subatomic particle reactions where the rest energy must be added to the balance.

From the two Friedmann equations we can show that

$$\frac{d}{dt}\left(\rho a^3\right) + p\frac{da^3}{dt} = 0 \tag{G12}$$

which is equivalent to the first law of thermodynamics derived in supplement F,

$$dU + dW = dQ \tag{G13}$$

where $dU = d(\rho V)$, $dQ = 0$, as expected for adiabatic expansion, and $dW = pdV$ is the work done by the system—or on the system if dW is negative. If we have a spherical region of space that is expanding with the rest of the universe ("comoving") and no heat is flowing in or out, a change in internal energy is compensated for by work being done on or by the system,

$$dU = -dW \tag{G14}$$

For nonrelativistic matter, $p \approx 0$ so $dW = 0$ and the mass-energy ρa^3 inside a^3 is a constant, from which it follows that $dU = 0$ and $\rho_m \sim a^{-3}$. For radiation, $p = \rho_r/3$, $\rho_r \sim a^{-4}$, and the mass-energy inside a^3 is not a constant. In this case, work must be done on or by the system. Note that the pressure can be negative, allowing ΔU to increase as the region expands. In any case, energy is still conserved.

In general, we can show from the Friedmann equations that

$$\frac{d\rho}{dt} = -3H\left(\rho + p\right) \tag{G15}$$

where

$$H = \frac{1}{a}\frac{da}{dt} \tag{G16}$$

is the Hubble parameter.

THE DENSITY OF THE UNIVERSE

From first of Friedmann's equations, the density of the Universe can be written

$$\rho = \frac{3}{8\pi Ga^2}\left[\left(\frac{da}{dt}\right)^2 + k\right] = \frac{3}{8\pi G}\left(H^2 + \frac{k}{a^2}\right) \tag{G17}$$

The critical density of the mass-energy of the Universe is obtained by setting $k = 0$:

$$\rho_c = \frac{3H^2}{8\pi G} \tag{G18}$$

With the current estimate of the Hubble parameter, $H = 1/13.8$ billion years, $\rho_c = 9.5 \times 10^{-30}$ g/cm^3 = 5.3×10^{-6} GeV/cm^3. If the average density of the Universe has this value, then space is flat; that is, the geometry of space is, on average, Euclidean. Of course, space is not Euclidean in the vicinity of massive objects such as stars. A flat universe is open but just on the verge of being closed. Note, however, that the same result will be obtained for $k = \pm1$, provided $H \gg 1/a$, or $da/dt \gg 1$ (da/dt is dimensionless in units $c = 1$).

Recall that the quantity k in Friedmann's equation is a constant of motion. This does not mean that ρ is constant, however, since H can in principle vary with time. In fact, we can show from the Friedmann equations that, for any k,

$$\frac{d\rho}{dt} = \frac{3H}{4\pi G}\frac{dH}{dt} \tag{G19}$$

GRAVITATIONAL REPULSION

From the second Friedmann equation

$$\frac{d^2a}{dt^2} = -\frac{4\pi G}{3}\left(\rho + 3p\right)a(t) \tag{G20}$$

we can see that general relativity allows for a gravitational repulsion if the pressure of the gravitating medium is sufficiently negative, $p < -\rho/3$.

Einstein noted that his equation for space-time curvature allowed for the inclusion of a constant term; this term modifies the Friedmann equation to read

$$\frac{1}{a}\frac{d^2a}{dt^2} = -\frac{4\pi G}{3}(\rho + 3p) + \frac{\Lambda}{3} \qquad \text{(G21)}$$

where Λ is called the cosmological constant. This term is repulsive when $\Lambda > 0$.

The cosmological term does not correspond to any physical medium but the curvature of space-time itself. Empty space, $\rho = p = 0$, still can have curvature when Λ is not zero. This solution of general relativity is called the *de Sitter universe*.

Let

$$\frac{\Lambda}{3} \equiv -\frac{4\pi G}{3}(\rho_v + 3p_v) \qquad \text{(G22)}$$

where the subscript v refers to "vacuum." In a vacuum, the density ρ and pressure p of matter and radiation are both zero. In this case,

$$\frac{\Lambda}{3} = \frac{1}{a}\frac{d^2a}{dt^2} = -\frac{4\pi G}{3}(\rho_v + 3p_v) = -\frac{4\pi G}{3}(1 + 3w_v)\rho_v \qquad \text{(G23)}$$

where

$$p_v = w_v \rho_v \qquad \text{(G24)}$$

is the equation of state for the "medium" that corresponds to the cosmological constant vacuum.

Now, since Λ is a constant,

$$\frac{d\rho_v}{dt} = -3H(\rho_v + p_v) = -3H(1 + w_v)\rho_v = 0 \qquad \text{(G25)}$$

so $w_v = -1$ and the cosmological constant is equivalent to a medium with constant mass density

$$\rho_v = \frac{\Lambda}{8\pi G} \qquad \text{(G26)}$$

and equation of state

$$p_v = -\rho_v \tag{G27}$$

with a positive acceleration (repulsive gravity),

$$\frac{1}{a}\frac{d^2a}{dt^2} = +\frac{8\pi G}{3}\rho_v \tag{G28}$$

COSMIC ACCELERATION VIA THE COSMOLOGICAL CONSTANT

Let us look at the acceleration that occurs by means of the cosmological constant. As we saw above, in a universe empty of matter and radiation, $\rho = p = 0$, we can still have

$$\frac{1}{a}\frac{d^2a}{dt^2} = \frac{8\pi G}{3}\rho_v = \frac{\Lambda}{3} \tag{G29}$$

where Λ is the cosmological constant. The solution is simple:

$$a(t) = A\exp(Ht) + B\exp(-Ht) \tag{G30}$$

where

$$H = \left(\frac{\Lambda}{3}\right)^{1/2} \tag{G31}$$

and A and B are constants. In most presentations, the B term is neglected since it rapidly goes to zero as t increases. However, note that the B term is significant and the A term is negligible when $t < 0$. There is no fundamental reason why these equations should not apply equally well for $t < 0$ as for $t > 0$.

For $t > 0$ we can drop B and make $A = a(0)$. Then

$$a(t) = a(0)\exp(Ht) \tag{G32}$$

Note that the rate of expansion also increases exponentially,

$$\frac{da}{dt} = a(0)H\exp\left(Ht\right) = Ha(t) \tag{G33}$$

More generally, in a universe including matter and radiation,

$$\frac{1}{a}\frac{d^2a}{dt^2} = -\frac{8\pi G}{3}\left(\frac{1}{2}\rho_m + \rho_r - \rho_v\right) \tag{G34}$$

and a net acceleration will occur when

$$\rho_v > \rho_r + \frac{1}{2}\rho_m \tag{G35}$$

COSMIC ACCELERATION VIA A SCALAR FIELD

Another way to achieve gravitational repulsion is for the dark energy to comprise a physical medium q, where $p_v \neq -\rho_v$ in general. In that case the density need not be constant and may evolve along with the other components, radiation and matter, as the Universe expands. This still unidentified substance has been dubbed *quintessence* and would currently constitute 70 percent of the mass-energy of the Universe.

In this case, the acceleration of the universe is given by

$$\frac{1}{a}\frac{d^2a}{dt^2} = -\frac{8\pi G}{3}\left[\frac{1}{2}\rho_m + \rho_r + \left(1 + 3w_q\right)\rho_q\right] \tag{G36}$$

where $w_q = p_q/\rho_q$. For a net repulsion,

$$w_q < -\frac{1}{3}\left(1 + \frac{\rho_m}{2\rho_q} + \frac{\rho_r}{\rho_q}\right) \tag{G37}$$

Currently, ρ_r is negligible and $\rho_m/\rho_q = 3/7$, so $w_q < -0.36$. In a universe dominated by dark energy, $w_q < -1/3$.

Let us consider the case where some kind of material "stuff" constitutes the field. In quantum field theory, that stuff will be composed of particles—the quanta of the field. In the simplest case, the field associated with ρ_q will be single valued—what is called a *scalar field*, designated by ϕ. (By contrast, the electric field E has three components and is, thus, a *vector field*.) As we will see, we can regard ϕ as a quantum field with an associated spin-0 quantum particle. In this picture, ϕ is viewed as another component to the Universe besides matter and radiation—one with negative pressure that results in a repulsion between particles.

As long as $p_q \neq -\rho_q$, the density can vary with time. This could solve the cosmological constant problem mentioned earlier. That is, the cosmological constant may be exactly zero, with the gravitational repulsion accounted for by the scalar field.

In classical field theory, it can be shown that the energy density of a scalar field is mathematically equivalent to the density of a unit mass non-relativistic particle moving in one dimension with coordinate ϕ. The sum of kinetic and potential energy densities is

$$\rho = \frac{1}{2}\left(\frac{d\phi}{dt}\right)^2 + u(\phi) \tag{G38}$$

The corresponding Lagrangian density will be

$$\mathcal{L} = \frac{1}{2}\left(\frac{d\phi}{dt}\right)^2 - u(\phi) \tag{G39}$$

Thus, in a sphere of radius a of uniform density, the Lagrangian is

$$L = \frac{4\pi}{3}a^3\mathcal{L} \tag{G40}$$

Lagrange's equation

$$\frac{d}{dt}\left[\frac{\partial L}{\partial\left(\frac{\partial\phi}{\partial t}\right)}\right] - \frac{\partial L}{\partial\phi} = 0 \tag{G41}$$

then gives

$$\frac{4\pi}{3}a^3\frac{d^2\phi}{dt^2} + 4\pi a^2\frac{da}{dt}\frac{d\phi}{dt} + \frac{4\pi}{3}a^3\frac{\partial u}{\partial\phi} = 0 \tag{G42}$$

or

$$\frac{d^2\phi}{dt^2} + 3H\frac{d\phi}{dt} + \frac{\partial u}{\partial\phi} = 0 \tag{G43}$$

for the equation of motion of ϕ. The second term is a "friction" or damping term that results from the expansion of the Universe, where

$$H = \frac{1}{a}\frac{da}{dt} \tag{G44}$$

is the Hubble parameter.

In terms of the density,

$$H^2 = \frac{8\pi}{3}\rho - \frac{k}{a^2} \tag{G45}$$

in Planck units ($G = 1$). Eliminating H from the equation of motion gives

$$\frac{d^2\phi}{dt^2} + 3\left(\frac{8\pi}{3} - \frac{k}{a^2}\right)^{1/2}\phi + \frac{\partial u}{\partial \phi} = 0 \tag{G46}$$

For any given $u(\phi)$, we can numerically integrate this to get $\phi(t)$, $\rho(t)$, $H(t)$, and $a(t)$.

Thus, if the Universe is assumed to be homogeneous and isotropic and dominated by a scalar potential, then its history for any assumed $u(\phi)$ is fully computable, depending only on the assumed initial values of ϕ, its time derivative, and the parameters of the model.

AT THE PLANCK SCALE

If we extrapolate the expanding Universe backward in time, we find we cannot go all the way to zero time. The smallest operationally definable time interval is the *Planck time*, which is 1.61×10^{-43} second. Likewise, the smallest operationally definable distance is that traveled by light in the Planck time is 4.05×10^{-35} meter, which is called the *Planck length*.

Let us see why this is the case. In order to measure a time t we need a clock with an uncertainty $\Delta t < t$. The Heisenberg time-energy uncertainty principle says that the product of Δt and the uncertainty in a measurement of energy in that time interval, ΔE, must be

$$\Delta E \Delta t \geq \hbar \tag{G47}$$

The precise statement of the uncertainty principle is $\Delta E \Delta t \geq \dfrac{\hbar}{2}$, where ΔE and Δt are statistical standard deviations; we need not worry about factors of two, which only complicate matters without changing the main conclusions.[1] Thus,

$$\Delta E \geq \frac{\hbar}{\Delta t} \geq \frac{\hbar c}{L} \tag{G48}$$

where $L = ct$ is the distance light goes in this time interval. The energy uncertainty is equivalent to the rest energy of a body of mass M,

$$\Delta E = Mc^2 \tag{G49}$$

Let L be the radius of a sphere. It follows that within a spherical region of space of radius L we cannot determine, by any measurement, that it contains a mass less than

$$\Delta E \geq \frac{\hbar}{\Delta t} \geq \frac{\hbar c}{L} \tag{G50}$$

A spherical body of mass M and radius

$$R = \frac{GM}{c^2} \tag{G51}$$

will be a black hole. Let us define

$$L_{PL} \equiv R = \left(\frac{\hbar G}{c^3} \right)^{1/2} \tag{G52}$$

which is the *Planck length*, $L_{PL} = 1.6 \times 10^{-35}$ meter. We can see that it represents the smallest length that can be operationally defined, that is, defined in terms of measurements that can be made with clocks and other instruments. If we tried to measure a smaller distance, the time interval would be smaller, the uncertainty in rest energy would be larger, the uncertainty in mass would be larger, and the region of space would be experimentally indistinguishable from a black hole. Since nothing inside a black hole can climb outside its gravitational field, we cannot see inside and thus cannot make a smaller measurement of distance.

Similarly, we can make no smaller measurement of time than the Planck time,

$$t_{PL} = \frac{L_{PL}}{c} = \left(\frac{\hbar G}{c^5} \right)^{1/2} \tag{G53}$$

which has the value $t_{PL} = 5.4 \times 10^{-44}$ second. Also of some interest are the Planck mass,

$$M_{PL} = \frac{\hbar}{cL_{PL}} = \left(\frac{\hbar c}{G} \right)^{1/2} \tag{G54}$$

which has a value of 2.2×10^8 kilograms, and the Planck energy,

$$E_{PL} = M_{PL} c^2 = \left(\frac{\hbar c^5}{G} \right)^{1/2} \tag{G55}$$

which has a value of 2.0×10^9 joules or 1.2×0^{19} GeV. These represent the uncertainties in rest mass and rest energy within the space of a Planck sphere or within a time interval equal to the Planck time. Note, however, that unlike the Planck length and time, the Planck mass and energy are not the smallest values possible for these quantities. In fact, they are quite large by subatomic standards.

Although $t = 0$ is specified as the time at which our Universe began, the smallest operationally defined time interval is the Planck time, 10^{-43} second. Within a time interval equal to the Planck time, Δt, the Heisenberg uncertainty principle says that the energy is uncertain by an amount on the order of $\hbar/\Delta t = E_{PL}$ (10^{19} GeV). That energy will be spread throughout a sphere with a radius on the order of the Planck length, 10^{-33} centimeter. The energy density within that sphere, for this brief moment, will be on the order of $(10^{19} \text{ GeV})/(10^{-33}\text{cm})^3 = 10^{118} \text{ GeV/cm}^3$, which we can call the Planck density, ρ_{PL}. This density would be equivalent to a cosmological constant,

$$\Lambda = 8\pi G \rho_{PL} \tag{G56}$$

Also note that the space and time intervals defined by the Planck scale apply to every momentum in time and position in space, not just the origin of the Universe. No measurements can be made with greater precision, no matter where or when.

In order to make the equations simpler, we can use "natural" units, where $\hbar = c = 1$. Then

$$L_{PL} = t_{PL} = \sqrt{G} \tag{G57}$$

$$E_{PL} = M_{PL} = \frac{1}{\sqrt{G}} \qquad (G58)$$

THE ENTROPY OF THE UNIVERSE

The entropy of a system of particles can be crudely approximated by the number of particles in the system.[2] Recall from supplement F that the entropy of a system containing Ω equally likely available states is

$$S = \log_2 \Omega \qquad (G59)$$

bits. For a system of N particles, each of which has n accessible states, we can write $\Omega = n^N$ so

$$S = N \log_2 n \qquad (G60)$$

Since n is not a very large number, its logarithm will be of order unity. So, it becomes a rough approximation, say to an order of magnitude, to write

$$S = N \qquad (G61)$$

The maximum entropy of a system will equal the maximum number of localizable particles within that system. This maximum results from the fact that particles cannot be localized within a region with dimensions less than the *Compton wavelength* that is associated with a particle in quantum mechanics. That wavelength is inversely proportional to the particle's energy, which gives a minimum-energy particle that can be found in the volume. A particle of energy ε has an associated Compton wavelength given by

$$\lambda = \frac{hc}{\varepsilon} \qquad (G62)$$

where h is Planck's constant. Suppose the particle is confined to a spherical volume of radius R. Then it will lose its identity when λ is on the order of $2\pi R$, the circumference of the sphere. It follows that the minimum energy a particle can have inside a sphere of radius R is

$$\varepsilon_{min} = \frac{hc}{2\pi R} = \frac{\hbar c}{R} \tag{G63}$$

Thus, the maximum entropy of a sphere with total energy E, is

$$S_{max} = \frac{E}{\varepsilon_{min}} \tag{G64}$$

Suppose the visible Universe is a sphere of radius R. It follows that the maximum entropy of the Universe is

$$S_{max} = \frac{ER}{\hbar c} = ER \tag{G65}$$

in units $\hbar = c = 1$.

What is the actual entropy of the Universe within our horizon today? Penrose argues that the entropy of black holes, which he estimates to be 10^{100}, should dominate the entropy of the Universe. The total energy of the visible Universe is at least 10^{80} GeV, based on the number of protons and other conventional particles. The current horizon is at $R = 1.2 \times 10^{26}$ meters. Thus the maximum entropy of the Universe, from (G65) is at least

$$S_{max} = \frac{ER}{\hbar c} = 10^{122} \tag{G66}$$

We can safely conclude that $S \ll S_{max}$ and no problem exists for the formation of local order, with at least 22 orders of magnitude of room available.

THE ENTROPY OF A BLACK HOLE

Suppose the rest energy of a spherical body of mass M and radius R equals its gravitational potential energy

$$M = \frac{GM^2}{R} \tag{G67}$$

where $c = 1$ and, since this is an order of magnitude calculation, we need not worry about factors of two or three. Then

$$R = GM \tag{G68}$$

This is half the *Schwarzschild radius*, below which a body is a black hole,

$$R_S = 2GM \tag{G69}$$

So, our body will be a black hole. Its maximum entropy will be, from (G65), on the order of

$$S_{BH} = MR = \frac{R^2}{G} = \left(\frac{R}{L_{PL}} \right)^2 \tag{G70}$$

Alternatively, we can write this as

$$S_{BH} = \left(\frac{M}{M_{PL}} \right)^2 \tag{G71}$$

Let us assume that at the Planck time our Universe was a sphere of radius equal to the Planck length. Because of the uncertainty principle, the energy uncertainty within this volume will be on the order of the Planck energy and the mass uncertainty on the order of the Planck mass. The maximum entropy of a sphere of radius R and mass M, is, from (G65) with $E = M$, on the order of

$$S_{max} = MR \tag{G72}$$

At the Planck scale, $M = M_{PL} = 1 / L_{PL}$ and $R = L_{PL}$, so $S_{max} = 1$. When a sphere of radius R is a black hole, its entropy is, from (G70),

$$S_{BH} = \left(\frac{R}{L_{PL}} \right)^2 \tag{G73}$$

so, at the Planck scale, $S_{max} = S_{BH} = 1$. That is, a Planck sphere is indistinguishable from a black hole, which, in turn, implies that the entropy of the Universe was maximal at that time. The entropy was both as high and as low as it could be!

   We can thus conclude that our Universe can have begun in total chaos, with no order or organization. If any order existed before that time, it was swallowed up by the primordial black hole.

We thus have the picture of the Universe starting out at the Planck time as a black hole of maximum entropy. It then explodes into an expanding gas of particles. Since the Universe is no longer a black hole, its entropy is then less than the maximum value it can have. It will not increase faster than the entropy of a black hole of tha same size. Thus, although it started out with maximum entropy, as it expands it finds increasing room for order to form. This resolves the apparent paradox in which the Universe begins in maximum chaos, with no structure or organization, and yet can have order form without any violation of the second law of thermodynamics.

INFLATION

I will not discuss all the various models of inflation that have been proposed, which can be found in many books and papers. Instead, I will limit myself to a particularly simple and elegant model proposed by André Linde called *chaotic inflation*.[3] Although Linde has also considered various forms of the potential within the framework of chaotic inflation, the simplest is,

$$u(\phi) = \frac{1}{2}m^2\phi^2 \qquad (G74)$$

where m is the only parameter. This is just the potential of a harmonic oscillator and m will be the mass of the quantum of the corresponding field. As before, the equation of motion includes damping,

$$\frac{d^2\phi}{dt^2} + 3H\frac{d\phi}{dt} + m^2\phi = 0 \qquad (G75)$$

where

$$H^2 = \frac{8\pi}{3}\rho - \frac{k}{a^2} = \frac{4\pi}{3}\left[\left(\frac{d\phi}{dt}\right)^2 + m^2\phi^2\right] - \frac{k}{a^2} \qquad (G76)$$

so,

$$\frac{d^2\phi}{dt^2} + 3\left\{\frac{4\pi}{3}\left[\left(\frac{d\phi}{dt}\right)^2 + m^2\phi^2\right] - \frac{k}{a^2}\right\}^{1/2}\frac{d\phi}{dt} + m^2\phi = 0 \qquad (G77)$$

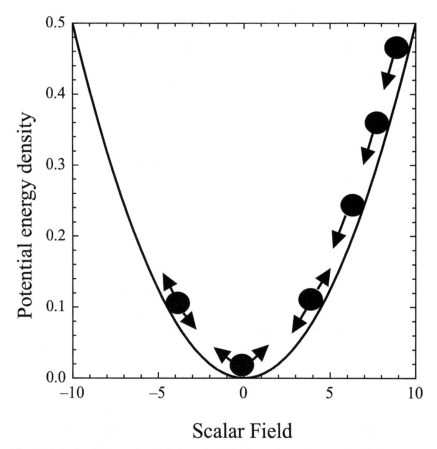

Scalar Field

Fig. G7.1. An illustration of chaotic inflation. Assume a potential energy density given by $u(\phi) = m^2\phi^2/2$ where ϕ is the scalar field. A quantum fluctuation produces $\phi = 10$ Planck units. The Universe rolls down the potential energy density hill, its volume increasing quasi exponentially. At the bottom it oscillates back and forth, producing particles, and eventually settles back to zero energy and zero field.

This can be solved numerically to get the time history of the early Universe. In the chaotic scenario, the Universe starts in a perfect vacuum, with $\phi = 0$. A quantum fluctuation sends it up the side of the potential, like the pendulum bob of an oscillator, and inflation is initiated (see fig. G7.1). Because of the damping, the Universe returns slowly to equilibrium and inflation is able to continue for a significant time.

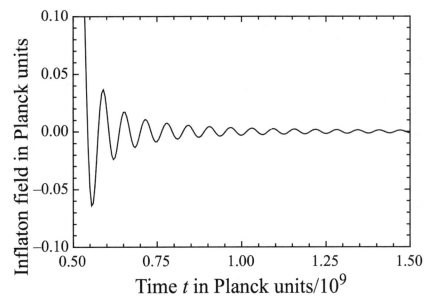

Fig. G7.2. The evolution of the Inflaton field $\phi(t)$ with time for chaotic inflation, where $m = 10^{-7}$ Planck mass and $k = 1$. From an initial true vacuum, a quantum fluctuation produces a scalar field equal to 10 Planck units. Inflation occurs as the Universe rolls down the potential hill, as illustrated in fig. G7.1. When it reaches the bottom it oscillates before coming to rest with zero scalar field. The region $t < 0.5$ is not shown.

Normally, when discussing inflation, one considers the completely geometrically flat universe, $k = 0$. However, in order to be consistent with some ideas that will be discussed in the following chapter, I will take $k = 1$, that is, assume a closed universe. As we will see, inflation still can occur under that condition.

I have taken the mass parameter m to be 10^{-7} Planck units, or on the order of 10^{12} GeV, a value that Linde says gives the necessary fluctuations for galaxy formation. For the purposes of illustration, I assume that the initial value of ϕ is 10 Planck units. Fig. G7.2 shows the evolution of ϕ with time. The zeros are suppressed on both axes so that the oscillations of ϕ at the end of evolution can be more readily seen. For $t < 0.5 \times 10^9$ Planck units, ϕ drops linearly from its initial value of 10.

The scale factor of the Universe will evolve with time according to

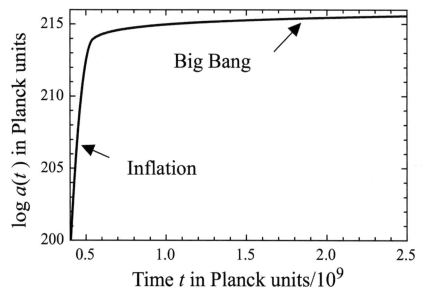

Fig. G7.3. The evolution of the scale factor of the Universe with time for chaotic inflation with $m = 10^{-7}$ Planck mass, $k = 1$, and an initial inflaton field of 10 Planck units. Exponential inflation of 214 orders of magnitude is followed by the beginning of the conventional big bang. This is illustrative only and not meant to exactly model our Universe, where the inflationary period was probably greater and the expansion probably included many more orders of magnitude. The time scale is approximately seconds $\times 10^{34}$.

$$\left(\frac{da}{dt}\right)^2 = \frac{a^2}{a_o^2} - 1 \tag{G78}$$

where

$$a_o = \left(\frac{3}{8\pi\rho}\right)^{1/2} \tag{G79}$$

When ρ is a constant this has the solution

$$a(t) = a_o \cosh(t / a_o) \tag{G80}$$

This implies that the Universe starts with $a = a_o$, which depends on the initial energy density. However, since the density also evolves as ϕ evolves, after the first few moments we must obtain $a(t)$, by numerically integrating

$$da = Hadt \qquad\qquad (G81)$$

The result, with the parameters used for illustration here, is shown in fig. G7.3. Again the zeros have been suppressed to provide a better illustration. Inflation occurs during the first 5×10^8 Planck times, or about 10^{-34} second, during which time the scale factor expanded by 214 orders of magnitude or the volume of the Universe increased by 642 orders of magnitude. It must be emphasized that this does not represent that exact situation that occurred in the actual early Universe, since the expansion factor depends very sensitively on the initial value of ϕ and the mass parameter m. In any case, when inflation ceases the much slower expansion of the big bang takes over.

NOTES

1. You will sometimes read that the energy-time uncertainty principle is invalid. However, if you simply think relativistically, where energy and time are the fourth components of the momentum and position 4-vectors, it should be obvious that if the normal momentum-position uncertainty principle is valid, then so too should be the energy-time principle.

2. Steven Frautschi, "Entropy in an Expanding Universe" in *Entropy, Information, and Evolution*, ed. Bruce H. Weber, David J. Depew, and James D. Smith, (Cambridge: MIT Press, 1988), pp. 11–22.

3. Roger Penrose, *The Emperor's New Mind: Concerning Computers, Minds, and the Laws of Physics* (Oxford: Oxford University Press, 1989), p. 342.

4. André Linde, "A New Inflationary Universe Scenario: A Possible Solution of the Horizon, Flatness, Homogeneity, Isotropy, and Primordial Monopole Problems," *Physics Letters* 108B (1982): 389–92.

MATHEMATICAL SUPPLEMENT H

PHYSICS OF THE VACUUM

THE PURE RADIATION FIELD

The classical electromagnetic field at a point in space where the charge and current density are zero can be derived from the vector potential wave equation

$$\nabla^2 \mathbf{A} - \ddot{\mathbf{A}} = 0 \tag{H1}$$

in units $c = 1$, where we can work in a gauge where the scalar potential $\phi = 0$. Imposing periodic boundary conditions, one solution is in the form of the plane wave,

$$\mathbf{A} = A_1 \hat{\mathbf{u}} e^{i(\mathbf{k}\bullet\mathbf{r}-\omega t)} + A_2 \hat{\mathbf{u}} e^{-i(\mathbf{k}\bullet\mathbf{r}-\omega t)} \tag{H2}$$

where $k^2 = \omega^2$ and the wave propagates in the direction of $\mathbf{k}$. Since $\nabla \bullet \mathbf{A} = 0$ in this gauge, $\mathbf{k} \bullet \hat{\mathbf{u}} = 0$.

More generally, we have a linear combination,

$$\mathbf{A} = \sum_\lambda \left(q_\lambda(t)\mathbf{A}_\lambda(\mathbf{r}) + q_\lambda^*(t)\mathbf{A}_\lambda^*(\mathbf{r}) \right) \tag{H3}$$

where

$$q_\lambda(t) = e^{-i\omega_\lambda t} \tag{H4}$$

Let us introduce the real variable

$$Q_\lambda = q_\lambda + q_\lambda^* = 2\cos(\omega_\lambda t) \tag{H5}$$

Note that

$$\ddot{Q}_\lambda + \omega_\lambda^2 Q_\lambda = 0 \tag{H6}$$

which is mathematically equivalent to a unit mass nonrelativistic harmonic oscillator, with the Hamiltonian,

$$H_\lambda = \frac{1}{2}\left(P_\lambda^2 + \omega_\lambda^2 Q_\lambda^2\right) \tag{H7}$$

where

$$\frac{\partial H_\lambda}{\partial Q_\lambda} = -\dot{P}_\lambda \quad \frac{\partial H_\lambda}{\partial P_\lambda} = \dot{Q}_\lambda = P_\lambda \tag{H8}$$

The total energy of the radiation field will then be

$$E = \sum_\lambda H_\lambda \tag{H9}$$

We can quantize the field by simply applying the results from the quantum harmonic oscillator. Let us drop the λ subscript, keeping in mind, then, that we are talking about a single oscillator with a given natural frequency ω and, thus, a radiation field of a single frequency equal to this value. Let us also work in units $\hbar = 1$ and recall that the mass of the oscillator is $m = 1$. Following the procedure found in quantum mechanics textbooks,[1] we define the operators

$$a = \sqrt{\frac{1}{2\omega}}\left(\omega Q + iP\right)$$
$$a^\dagger = \sqrt{\frac{1}{2\omega}}\left(\omega Q - iP\right) \tag{H10}$$

Note that

$$\left[a, a^\dagger\right] = \left[a^\dagger, a\right] = 1 \quad \left[a, a\right] = \left[a^\dagger, a^\dagger\right] = 0 \tag{H11}$$

Define the *number operator*

$$N = a^\dagger a \tag{H12}$$

A bit of algebra will show that

$$H = \left(N + \tfrac{1}{2}\right)\omega \tag{H13}$$

The eigenvalue equation for N is

$$N\left|n\right\rangle = n\left|n\right\rangle \tag{H14}$$

so that

$$H\left|n\right\rangle = E_n\left|n\right\rangle = \left(n + \tfrac{1}{2}\right)\omega\left|n\right\rangle \tag{H15}$$

and the energy eigenvalues then are

$$E_n = \left(n + \tfrac{1}{2}\right)\omega \tag{H16}$$

Again, I refer you to the textbooks to show that

$$a\left|n\right\rangle = \sqrt{n}\left|n-1\right\rangle$$
$$a^\dagger\left|n\right\rangle = \sqrt{n+1}\left|n+1\right\rangle \tag{H17}$$

so we see that a and $a^\dagger$ are *annihilation* and *creation operators*, respectively.

The physical interpretation in terms of the pure radiation field is as follows. The energy levels of the harmonic oscillator are equally spaced like the rungs on a ladder by an amount ω, as illustrated in fig. 8.1 (p. 144) ($h\nu = \omega$), where the energy of a level is given by

$$E_n = \left(n + \tfrac{1}{2}\right)\omega \tag{H18}$$

Each energy level contains n photons (or "quanta"), each of energy ω. The annihilation operator a, operating on the energy eigenstate $\left|n\right\rangle$, takes us from one rung on the ladder to the one below by removing one photon.

The creation operator $a^\dagger$, operating on the energy eigenstate $|n\rangle$, takes us from one rung on the ladder to the one above by adding one photon.

When we get to the bottom rung, the state $|0\rangle$ with zero photons, we do not have zero energy. Rather, we have a state with energy

$$E_o = \tfrac{1}{2}\omega \quad \text{(bosons)} \tag{H19}$$

This is called the *zero-point energy.*

It can be shown that this will be zero-point energy for all bosons, such as spin-0 particles. The zero-point energy for fermions, on the other hand, is

$$E_o = -\tfrac{1}{2}\omega \quad \text{(fermions)} \tag{H20}$$

ZERO-POINT ENERGY FROM THE UNCERTAINTY PRINCIPLE

The zero-point energy of a *material* harmonic oscillator can be understood crudely in terms of the uncertainty principle. Let us consider a one-dimensional nonrelativistic oscillator of mass m. Suppose it has a momentum Δp and displacement from equilibrium Δx given by the minimum allowed by the uncertainty principle,

$$\Delta p \Delta x = \frac{1}{2} \tag{H21}$$

where we work in units $\hbar = 1$. The oscillator energy will be

$$\begin{aligned} E &= \frac{(\Delta p)^2}{2m} + \frac{1}{2}m\omega^2(\Delta x)^2 \\ &= \frac{1}{8m(\Delta x)^2} + \frac{1}{2}m\omega^2(\Delta x)^2 \end{aligned} \tag{H22}$$

Differentiating this with respect to Δx and setting the result equal to zero, we get the value of Δx for which E is minimum,

$$\Delta x = \sqrt{\frac{1}{2m\omega}} \tag{H23}$$

From (H22) that minimum is, as was independently determined above,

$$E_o = \tfrac{1}{2}\omega \tag{H24}$$

This basically says that an oscillator can never be at rest.

I know of no similarly simple explanation that can be given for the negative zero-point energy of fermions.

VACUUM ENERGY

The vacuum is conventionally defined as the state $|0\rangle$, the state of zero particles. Since, in quantum field theory, fields are always associated with particles, their quanta, the vacuum also has zero fields. If we start with a boson field and remove all the bosons, we are left with a state with zero bosons but nonzero energy. This energy seems to be infinite or at least very large. To see this, note that the number of boson states in a volume of phase space $d^3k\,d^3x$ is

$$dN = \frac{g}{h^3}\,d^3k\,d^3x \tag{H25}$$

where h is Planck's constant, and the factor g is the number of spin states and other degrees of freedom. Assuming all directions of the 3-vector $\mathbf{k}$ to be equally likely, we can integrate over its solid angle Ω, where $d^3k = k^2\,dk\,d\Omega$, to get, with $\hbar = 1$ so that $h = 2\pi$, the density of states defined as the number of states per unit configuration space volume,

$$dn = \frac{4\pi g}{\left(2\pi\right)^3}\,k^2\,dk \tag{H26}$$

The total zero-point energy density for bosons then is

$$\rho_b = \frac{1}{2}\int_0^\infty \omega\,dn = \frac{4\pi g}{\left(2\pi\right)^3}\frac{1}{2}\int_0^\infty \omega k^2\,dk \tag{H27}$$

where $\omega = (k^2 + m^2)^{1/2}$ for scalar bosons of mass m, and $\omega = k$ for photons and other massless vector bosons. In either case, the integral is infinite.

However, since $k = 2\pi/\lambda$, where λ is the de Broglie-Compton wavelength, we should have a maximum k when λ is on the order of the Planck length, smaller distances being operationally indefinable. That is, $k_{max} = M_{PL} = 10^{19}$ GeV. Then the integral has a cutoff,

$$\rho_b = \frac{4\pi g}{(2\pi)^3} \frac{1}{2} \int_0^{M_{PL}} \omega k^2 dk$$

$$= \frac{4\pi}{(2\pi)^3} \frac{1}{2} g \frac{M_{PL}^4}{4}$$

(H28)

Putting in numbers, taking g of order unity, $\rho_b \approx 10^2 (10^{19} \text{ GeV})^4 = 10^{74}$ GeV$^4 = 10^{115}$ GeV/cm^3, not infinite but very large, nonetheless.

In general relativity the absolute magnitude of the momentum-energy tensor determines the curvature of space-time, and any vacuum energy ρ_v is equivalent to a cosmological constant,

$$\Lambda = 8\pi G \rho_v$$

(H29)

We estimate from observations that ρ_v is 120 orders of magnitude lower than ρ_b. This is called the *vacuum energy* or the *cosmological constant problem*.

DON'T FORGET THE FERMION ENERGY

Let us define

$$C = \frac{4\pi}{(2\pi)^3} \frac{1}{2} \int_0^{k_{max}} \omega k^2 dk$$

(H30)

which, as we saw above, has the value 10^{74} GeV$^4 = 10^{115}$ GeV/cm^3 for $k_{max} = M_{PL}$. Then the total vacuum energy density of boson and fermion fields is

$$\rho_b + \rho_f = C \left(\sum_b g_b - 2 \sum_f g_f \right)$$

(H31)

where the factor of 2 for the fermions results from the fact that we must also count antifermions. The two sum terms are just the total number of spin states and the other "internal" degrees of freedom for bosons and fermions. respectively. They will cancel exactly for perfect supersymmetry, which may have been the situation in the very early Universe.

DON'T FORGET THE NEGATIVE ENERGY STATES

The vacuum energy problem might be solved by considering negative energy states.[2] I will present this argument in a different fashion than the references.

Recall above that we gave as a solution to the pure radiation field wave equation (H2)

$$\mathbf{A} = A_1 \hat{\mathbf{u}} e^{i(\mathbf{k} \cdot \mathbf{r} - \omega t)} + A_2 \hat{\mathbf{u}} e^{-i(\mathbf{k} \cdot \mathbf{r} - \omega t)} \tag{H32}$$

which corresponded to a plane wave propagating in the $\mathbf{k}$-direction. When this is quantized, the quanta are photons of energy ω and 3-momentum $\mathbf{k}$.

However, an equally valid solution is

$$\mathbf{A}' = A_1' \hat{\mathbf{u}} e^{i(\mathbf{k} \cdot \mathbf{r} + \omega t)} + A_2' \hat{\mathbf{u}} e^{-i(\mathbf{k} \cdot \mathbf{r} + \omega t)} \tag{H33}$$

which is a plane wave propagating in the $-\mathbf{k}$-direction. This solution usually rejected as unphysical.

Note, though, that we will get the same result as we did previously if we write $\omega = -\omega'$. This would correspond to photons with negative energy. The quantization process would be the same, only we would have energy levels

$$E'_n = \left(n + \tfrac{1}{2}\right)\omega' = -\left(n + \tfrac{1}{2}\right)\omega \tag{H34}$$

and a zero-point energy of

$$E_o = -\tfrac{1}{2}\omega \quad \text{(negative energy bosons)} \tag{H35}$$

Including both sets of solutions, as the mathematics requires, we have an exact cancellation of the zero-point energy, which would work for both bosons and fermions, and would apply even when supersymmetry is broken.

If negative energies seem unacceptable, try looking at them another way. The negative energies become positive when we reverse the arrow of time, while the previous positive energies become negative, so we still get the same cancellation of the zero-point energy. Since no fundamental arrow of time exists in physics, that is, no experiment distinguishes between the two time directions, we have no basis for tossing out either set of solutions. Including both sets gives zero vacuum energy.

THE LAWS OF THE VACUUM

A vacuum is obtained by taking a set of particles and removing all the particles. The mathematical procedure is implied by the models of free bosonic and fermionic fields described above. If $|n\rangle$ is the state vector representing a set of n particles, and a is the annihilation operator defined by

$$a|n\rangle = \sqrt{n}\,|n-1\rangle \tag{H36}$$

where the $\sqrt{n}$ is for normalization purposes, so that

$$\langle n|n\rangle = 1 \tag{H37}$$

for all n, then n-successive operations of the annihilation operator will take us to the state $|0\rangle$, which contains no particles.

Although the state $|0\rangle$ has no particles, it remains a legitimate object for our abstract theoretical description. It is just as well defined as $|1\rangle$ or $|137938511\rangle$. Thus any principles we apply to $|n\rangle$ can also be legitimately applied to $|0\rangle$. Using the techniques developed in supplement D, if

$$U \approx 1 + i\varepsilon G \tag{H38}$$

is an infinitesimal transformation, where G is the generator of the transformation, and if $|n\rangle$ is invariant under that transformation,

$$\langle n|U^{\dagger}U|n\rangle = \langle n|n\rangle = 1 \tag{H39}$$

then G is a hermitian operator representing an observable that is conserved.

QUANTUM TUNNELING

Consider a nonrelativistic particle of mass m and energy E incident on a square potential barrier of height $V > E$ and width b centered at $x = 0$, as shown in fig. 8.2 (p. 150). The wave function for the particle on the left of the barrier is

$$\psi(x,t) = A\exp\left[i(px - Et)\right] + B\exp\left[i(-px - Et)\right] \tag{H40}$$

for $x < -\dfrac{b}{2}$, where $p = (2mE)^{1/2}$ and the second term is the reflection

from the interface at $x = -b/2$.

Inside the barrier the wave function will be

$$\psi(x,t) = C\exp\left[i(px - Et)\right] + D\exp\left[i(-px - Et)\right] \tag{H41}$$

for $-\dfrac{b}{2} < x < \dfrac{b}{2}$, where $p = i\left[2m(V - E)\right]^{1/2}$. Here we have "nonphys-

ical" solutions in which the particle has imaginary momentum.

The wave function to the right of the barrier will be

$$\psi(x,t) = F\exp\left[i(px - Et)\right] \tag{H42}$$

for $x > \dfrac{b}{2}$, where no particles come in from the right. These solutions

can be easily verified by substituting them into the Schrödinger equations,

$$-\frac{1}{2m}\frac{\partial^2 \psi}{\partial x^2} + V\psi = E\psi$$

(H43)

$$i\frac{\partial \psi}{\partial t} = E\psi$$

where $\hbar = 1$ and $V = 0$ outside the barrier.

We then have a double-boundary value problem. The probability for reflection by and transmission through the barrier is derived in many textbooks. Let me just present the result. The probability for transmission through the barrier to $x = +b/2$ is

$$P = \exp\left\{-2\left[2m(V-E)\right]^{1/2}b\right\}$$

(H44)

When the barrier has an arbitrary shape $V(x)$ from x_1 to x_2, we can obtain P from

$$P = \exp\left\{-2\int_{x_1}^{x_2}\left[2m(V(x)-E)\right]^{1/2}dx\right\}$$

(H45)

This is the same result as obtained in textbooks using the WKB approximation.

THE WAVE FUNCTION OF THE UNIVERSE

Consider the Lagrangian[3]

$$L = \frac{3\pi}{4G}\left[-\left(\frac{da}{dt}\right)^2 a + a\left(1-\frac{a^2}{a_o^2}\right)\right]$$

(H46)

where a is some coordinate. This looks rather arbitrary, but is chosen in hindsight. The canonical momentum conjugate to a is

$$p = \frac{\partial L}{\partial \left(\dfrac{da}{dt} \right)} = -\frac{3\pi}{2G} a \frac{da}{dt} \tag{H47}$$

The Hamiltonian then is

$$H = p \frac{da}{dt} - L$$

$$= -\frac{3\pi}{4G} a \left[1 + \left(\frac{da}{dt} \right)^2 - \frac{a^2}{a_o^2} \right] \tag{H48}$$

Consider the case where $H = 0$. Then

$$1 + \left(\frac{da}{dt} \right)^2 - \frac{a^2}{a_o^2} = 0 \tag{H49}$$

This is precisely one of the Friedmann equations discussed in supplement G:

$$\left(\frac{da}{dt} \right)^2 - \frac{8\pi\rho G}{3} a^2 = -k \tag{H50}$$

where a is the scale factor of the Universe, $a_o^2 = \dfrac{3}{8\pi\rho G}$ and $k = 1$, indicating a closed universe.

If the Universe is empty, then $\rho = \dfrac{\Lambda}{8\pi G}$ and $a_o^2 = \dfrac{3}{\Lambda}$ is constant. As

we saw, in this case, the scale factor varies with time as

$$a(t) = a_o \cosh \left(\frac{t}{a_o} \right) \tag{H51}$$

where the origin of our Universe is at $t = 0$. Nothing prevents us from applying this for earlier times. For $t < 0$ the Universe contracts, reaching a_o at $t = 0$. Thereafter it expands very rapidly and can provide for inflation. Note that the scale factor is never smaller than a_o.

The fact that $k = 1$ may seem to contradict the familiar prediction of inflationary cosmology that $k = 0$, that is, that the Universe is flat. Actually, a highly flat universe is not impossible with $k = 1$. Such a universe is simply huge, like a giant inflated balloon, with the part within our horizon simply a tiny patch on the surface. It eventually collapses in a "big crunch" but it will do this so far into the future that we can hardly speculate about it, except to conjecture that maximum entropy or "heat death" is likely to happen before the collapse is complete. Models for a closed universe exist that are consistent with all current data.[4] In fact, now that we know the universal expansion is accelerating, such models are even more viable.

Now, let us write the Friedmann equation

$$p^2 + \left(\frac{3\pi}{2G}\right)^2 a^2 \left(1 - \frac{a^2}{a_o^2}\right) = 0 \tag{H52}$$

This is a classical physics equation. One way to go from classical physics to quantum physics is by *canonical quantization*, in which we replace the

momentum p by an operator: $p = -i\dfrac{\partial}{\partial a}$,

$$\left[\frac{d^2}{da^2} - \left(\frac{3\pi}{2G}\right)^2 a^2 \left(1 - \frac{a^2}{a_o^2}\right)\right]\psi = 0 \tag{H53}$$

This is a special case of the *Wheeler-DeWitt equation*, where ψ is grandly referred to as the *wave function of the Universe*.

In general, the wave function of the Universe described by the Wheeler-DeWitt equation is a function of functions (that is, it is a *functional*) that specifies the geometry of the Universe at every point in three-dimensional space, with time puzzlingly absent as a parameter. Here we consider the situation where the geometry can be described by a single parameter, the radial scale factor a. This should be a reasonable approximation for the early Universe and greatly simplifies the problem. Aficionados of quantum gravity may object that this is too simplified, but since no general model of quantum gravity exists and all attempts at such a model involve mathematics well beyond the scope of this book,[5] we can at least consider this situation and hopefully gain some insight as to what might have gone on around the Planck time.

So, let us proceed with this approximation. We see that, mathematically speaking, the Wheeler-DeWitt equation in this case is simply the time-independent Schrödinger equation for a nonrelativistic particle of mass $m = 1/2$ in Planck units and zero total mechanical energy, moving in one dimension with coordinate a in a potential

$$V(a) = \left(\frac{3\pi}{2G}\right)^2 a^2 \left(1 - \frac{a^2}{a_o^2}\right) \tag{H54}$$

This potential is shown in fig. H8.1. Note that it includes the region $a < a_o$. The mathematical solution of the Friedmann equation for this region is

$$a(\tau) = a_o \cos\left(\frac{\tau}{a_o}\right) \tag{H55}$$

where τ is a real number and so the "time" $t = i\tau$ is an imaginary number. Thus the region $a < a_o$ cannot be described in terms of the familiar operational time, which is a real number read off a clock. Like the particle

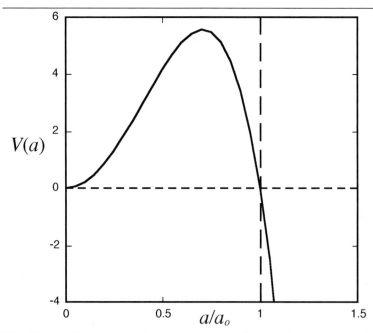

Fig. H8.1. The potential energy in the Wheeler-DeWitt equation. The vertical scale is in units of $\left(\dfrac{3\pi}{2G}\right)^2$.

under the square barrier described earlier, this is an "unphysical" region, meaning it is a region not amenable to observation. However, we saw that meaningful results can still be obtained when our equations are extended into the unphysical region. Using imaginary time here is analogous to using imaginary momentum for the square barrier case.

The potential energy in that region very much resembles the kind of potential barriers encountered, for example, in nuclear physics, where α-decay is explained as a quantum tunneling process of the kind described above. A variable barrier $V(x)$ in the range $a < x < b$ can be treated as a sequence of infinitesimal square barriers. The probability for transmission through the barrier will be, from (H45),

$$P \approx \exp\left(-2\int_a^b dx\, 2m\left|V(x) - E\right|^{1/2}\right) \tag{H56}$$

Here, $2m = 1$, $E = 0$, $a = 0$, and $b = a_o$, so

$$P \approx \exp\left(-2\int_o^{a_o} da\, K(a)\right) \tag{H57}$$

where

$$K(a) = \left(\frac{3\pi}{2G}\right)^2 a\left(\frac{a^2}{a_o^2} - 1\right)^{1/2} \tag{H58}$$

The above integral can be shown to yield,

$$P \approx \exp\left(-\frac{3}{8G^2\rho}\right) \tag{H59}$$

At the Planck time we can expect ρ to be on the order of the Planck density. In that case, the tunneling probability is $\exp(-3/8) = 68.7$ percent. This suggests that the unphysical region is highly unstable and that that 68.7 percent of all universes will be found in the physical state.

The region $a < a_o$ is a classically disallowed region that is allowed quantum mechanically and is described by a solution of the Wheeler-DeWitt equation. The particular solution will depend on boundary conditions. Hartle and Hawking propose equal amounts of incoming and outgoing waves in their "no boundary" model.

$$\psi_{HH}(a > a_o) = K(a)^{-1/2} \cos\left[\frac{\pi}{2} a_o^2 \left(\frac{a^2}{a_o^2} - 1\right)^{3/2}\right] \qquad \text{(H60)}$$

$$\psi_{HH}(a > a_o) = K(a)^{-1/2} \cos\left[\frac{\pi}{2} a_o^2 \left(\frac{a^2}{a_o^2} - 1\right)^{3/2}\right] \qquad \text{(H61)}$$

Vilenkin[6] proposes a boundary condition of outgoing waves only, which gives

$$\psi_V(a > a_o) = K(a)^{-1/2} \exp\left[-i\frac{\pi}{2} a_o^2 \left(\frac{a^2}{a_o^2} - 1\right)^{3/2}\right] \qquad \text{(H62)}$$

and

$$\psi_V(0 < a < a_o) = |K(a)|^{-1/2} \left\{ \frac{1}{2}\exp\left[-\frac{\pi}{2} a_o^2 \left(1 - \frac{a^2}{a_o^2}\right)^{3/2}\right] + i\exp\left[-\frac{\pi}{2} a_o^2 \left(1 - \frac{a^2}{a_o^2}\right)^{3/2}\right] \right\} \qquad \text{(H63)}$$

It is to be noted that these solutions have been obtained using WKB approximation and do not apply for the regions around $a = 0$ and $a = a_o$. The Hartle-Hawking wave function,[7] which is the simpler of the two, is shown in fig. H8.2, where I have extrapolated to zero at $a = 0$ and smoothly connected the inside and outside solutions at $a = ao$.

In Vilenkin's picture, the Universe is born by quantum tunneling from the classically unphysical region $a < a_o$ to the classically physical one $a > a_o$. Vilenkin construes the region $a < a_o$ as "nothing" since it does not contain classical space-time, explaining, "'Nothing' is the realm of unrestrained quantum gravity; it is a rather bizarre state in which all our basic notions of space, time, energy, entropy, etc., lose their meaning." While Vilenkin's association of the region inside the barrier as "nothing" is arguable, the region cannot be described in terms of a measurable time and thus a measurable space.

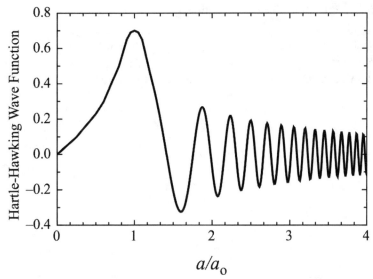

Fig. H8.2. The Hartle-Hawking wave function of the universe.

Atkatz and Pagels have shown that this tunneling process will only work for a closed universe, that is, one in which the cosmological parameter $k = 1$.[8] As we saw above, this does not contradict the inflationary model.

NOTES

1. For example, see J. J. Sakurai, *Modern Quantum Mechanics* (New York: Addison-Wesley, 1985), pp. 89–97.

2. Robert D. Klauber, "Mechanism for Vanishing Zero Point Energy," http://www.arxiv.org/pdf/astro-ph/0309679 (accessed January 20, 2005); J. W. Moffat, "Charge Conjugation Invariance of the Vacuum and the Cosmological Constant Problem," http:// xxx.lanl.gov/abs/hep-th/0507020m (accessed July 21, 2005).

3. The following development is based on David Atkatz, "Quantum Cosmology for Pedestrians," *American Journal of Physics* 62 (1994): 619–27.

4. Marc Kamionkowski and Nicolaos Toumbas, "A Low-Density Closed Universe," *Physical Review Letters* 77 (1996): 587–90.

5. Lee Smolin, *Three Roads to Quantum Gravity* (New York: Basic Books, 2001).

6. A. Vilenkin, "Birth of Inflationary Universes," *Physical Review* D27 (1983): 2848–55.

7. J. B. Hartle and S. W. Hawking, "Wave Function of the Universe," *Physical Review* D28 (1983): 2960–75.

8. Atkatz, "Quantum Cosmology for Pedestrians," pp. 619–27.

BIBLIOGRAPHY

Aspect, Alain, Phillipe Grangier, and Roger Gerard. "Experimental Realization of the Einstein-Podolsky-Rosen Gedankenexperiment: A New Violation of Bell's Inequalities." *Physical Review Letters* 49 (1982): 91–94.

———. "Experimental Tests of Bell's Inequalities Using Time-Varying Analyzers." *Physical Review Letters* 49 (1982): 1804–1809.

Atkatz, David. "Quantum Cosmology for Pedestrians." *American Journal of Physics* 62 (1994): 619–27.

Atkatz, David, and Heinz Pagels. "Origin of the Universe as Quantum Tunneling Event." *Physical Review* D25 (1982): 2065–73.

Barbour, J., and H. Pfister, eds. *Mach's Principle—From Newton's Bucket to Quantum Gravity*. Boston: Birkhauser, 1995.

Barrow, John D., and Frank J. Tipler. *The Anthropic Cosmological Principle*. Oxford: Oxford University Press, 1986.

Bell, John S. "On the Einstein-Podolsky-Rosen Paradox." *Physics* 1 (1964): 195–200.

Bohm, D. and B. J. Hiley. *The Undivided Universe: An Ontological Interpretation of Quantum Mechanics*. London: Routledge, 1993.

Bordag, M., U. Mohideen, and V. M. Mostepanenko. "New Developments in the Casimir Effect." *Physics Reports* 353 (2001): 1–205.

Boswell, James. *Life of Johnson.* 1791. Oxford: Oxford University Press, 1980.

Butterfield, J., and C. J. Isham, "Space-time and the Philosophical Challenge of Quantum Gravity." In *Physics Meets Philosophy at the Planck Scale*, edited by C. Callender and N. Huggett. Cambridge, UK: Cambridge University Press, 2001.

Byers, Nina. "E. Noether's Discovery of the Deep Connection Between Symmetries and Conservation Laws." *Israel Mathematical Conference Proceedings*, 12 (1999). http://www.physics.ucla.edu/~cwp/articles/noether.asg/noether.html (accessed November 5, 2004).

Casimir, H. B. G. "On the Attraction between Two Perfectly Conducting Plates." *Proceedings Koninkl. Ned. Akad. Wetenschap.* 51 (1948): 793–95.

Costa de Beauregard, Olivier. "Une réponse à l'argument dirigé par Einstein, Podolsky, et Rosen contre l'interpretation bohrienne de phénomènes quantiques." *Comptes Rendus* 236 (1953): 1632–34.

———. "Time, Symmetry, and the Einstein Paradox." *Il Nuovo Cimento* 42B1 (1977): 41–63.

Cramer, John G., "The Transactional Interpretation of Quantum Mechanics." *Reviews of Modern Physics* 58 (1986): 647–88.

Davies, Paul. *About Time: Einstein's Unfinished Revolution.* New York: Simon & Schuster, 1995.

Deutsch, David. "Quantum Theory of Probability and Decisions." *Proceedings of the Royal Society* A456 (2000): 1759–74.

Dicke, R. H. *Theoretical Significance of Experimental Relativity.* New York: Gordon & Breach, 1964.

Dirac, P. A. M. *The Principles of Quantum Mechanics.* Oxford: Oxford University Press, 1930.

Einstein, A., B. Podolsky, and N. Rosen. "Can the Quantum Mechanical Description of Physical Reality Be Considered Complete?" *Physical Review* 47 (1935): 777–80.

Ferris, Timothy. *The Whole Shebang: A State-of-the-Universe Report.* New York: Simon & Schuster, 1997.

Feynman, Richard P. *The Principle of Least Action in Quantum Mechanics.* Ann Arbor, MI: University Microfilms, 1942.

———. *QED: The Strange Theory of Light and Matter.* Princeton, NJ: Princeton University Press, 1985.

———."Space-Time Approach to Non-relativistic Quantum Mechanics." *Reviews of Modern Physics* 20 (1948): 367–87.

———. "Space-Time Approach to Quantum Electrodynamics." *Physical Review* 76 (1949): 769–89.

———. "The Theory of Positrons." *Physical Review* 76 (1949): 749–59.

Feynman, Richard P., and A. R. Hibbs. *Quantum Mechanics and Path Integrals.* New York: McGraw-Hill, 1965.

Frautschi, Steven. "Entropy in an Expanding Universe." In *Entropy, Information, and Evolution*, edited by Bruce H. Weber, David J. Depew, and James D. Smith, 11–22. Cambridge: MIT Press, 1988.

Gasperini M., and G. Veneziano. "The Pre–Big Bang Scenario in String Cosmology." *Physics Reports* 373 (2003): 1–212.

Goldstein, Herbert. *Classical Mechanics.* 2nd ed. New York: Addison-Wesley, 1980.

Greene, Brian. *The Elegant Universe: Superstrings, Hidden Dimensions, and the Quest for the Ultimate Theory.* New York: Norton, 1999.

Griffiths, David. *Introduction to Electrodynamics.* Upper Saddle River, NJ: Prentice-Hall, 1999.

———. *Introduction to Elementary Particles.* New York: Wiley, 1987.

Guth, Alan. "The Inflationary Universe: A Possible Solution to the Horizon and Flatness Problems." *Physical Review* D23 (1981): 347–56.

————. *The Inflationary Universe.* New York: Addison-Wesley, 1997.

Hartle, J. B., and S. W. Hawking. "Wave Function of the Universe." *Physical Review* D28 (1983): 2960–75.

Hawking, Stephen W. *A Brief History of Time: From the Big Bang to Black Holes.* New York: Bantam, 1988.

————. "Quantum Cosmology." In *Three Hundred Years of Graviation,* edited by S. W. Hawking and W. Israel, 631–51. Cambridge: Cambridge University Press, 1987.

Hoddeson, Lillian, Laurie Brown, Michael Riordan, and Max Dresden, eds. *The Rise of the Standard Model: Particle Physics in the 1960s and 1970s.* Cambridge: Cambridge University Press, 1997.

Hume, David. *An Enquiry concerning Human Understanding.* 1748. Edited by Tom L. Beauchamp. Oxford: Oxford University Press, 1999.

Jaffe, R. L. "The Casimir Effect and the Quantum Vacuum." http://arxiv.org/abs/hep-th/0503158 (accessed March 21, 2005).

Kamionkowski, Marc, and Nicolaos Toumbas. "A Low-Density Closed Universe." *Physical Review Letters* 77 (1996): 587–90.

Kane, Gordon. *The Particle Garden: Our Universe as Understood by Particle Physicists.* New York: Addison-Wesley, 1995.

————. *Supersymmetry: Unveiling the Ultimate Laws of Nature.* Cambridge, MA: Perseus, 2000.

Kazanas, D. "Dynamics of the Universe and Spontaneous Symmetry Breaking." *Astrophysical Journal* 241 (1980): L59–63.

Klauber, Robert D. "Mechanism for Vanishing Zero Point Energy." http://www.arxiv.org/pdf/astro-ph/0309679 (accessed January 20, 2005).

Kolb, Edward W., Sabino Matarasse, Alession Notari, and Antonio Riotto. "Primordial Inflation Explains Why the Universe Is Accelerating Today." http://arxiv.org/abs/hep-th/0503117 (accessed March 14, 2005).

Kuhn, Thomas. *The Structure of Scientific Revolutions.* Chicago: University of Chicago Press, 1970.

Lamoreaux, S. K. "Demonstration of the Casimir Force in the 0.6 to 6 µM Range." *Physical Review Letters* 78 (1997): 5–8; 81 (1997): 5475–76.

Lederman, Leon M., and Christopher T. Hill. *Symmetry and the Beautiful Universe.* Amherst, NY: Prometheus Books, 2004.

Linde, André. "A New Inflationary Universe Scenario: A Possible Solution of the Horizon, Flatness, Homogeneity, Isotropy, and Primordial Monopole Problems." *Physics Letters* 108B (1982): 389–92.

————. "Quantum Creation of the Inflationary Universe." *Lettere al Nuovo Cimento* 39 (1984): 401–405.

Mandl, F., and G. Shaw. *Quantum Field Theory.* Rev. ed. New York: Wiley, 1984.

Milne, E. A. *Relativity, Gravitation, and World Structure*. Oxford: Oxford University Press, 1935.

Moffat, J. W. "Charge Conjugation Invariance of the Vacuum and the Cosmological Constant Problem." July 20, 2005. http://xxx.lanl.gov/abs/hep-th/0507020 (accessed July 21, 2005).

Ostriker, Jeremiah P., and Paul J. Steinhardt. "The Quintessential Universe." *Scientific American* (January 2001): 46–53.

Pais, Abraham. *"Subtle Is the Lord . . .": The Science and the Life of Albert Einstein*. Oxford: Oxford University Press, 1982.

Penrose, Roger. *The Emperor's New Mind: Concerning Computers, Minds, and the Laws of Physics*. Oxford: Oxford University Press, 1989.

———. *The Road to Reality: A Complete Guide to the Laws of the Universe*. New York: Knopf, 2004.

Perlmutter, S., et al. "Measurements of Omega and Lambda from 42 High-Redshift Supernovae." *Astrophysical Journal* 517 (1999): 565–86.

Price, Huw. *Time's Arrow and Archimedes Point: New Directions for the Physics of Time*. Oxford: Oxford University Press, 1996.

Reif, F. *Fundamentals of Statistical and Thermal Physics*. New York: McGraw-Hill, 1965.

Reiss, A., et al. "Observational Evidence from Supernovae for an Accelerating Universe and a Cosmological Constant." *Astronomical Journal* 116 (1998): 1009–38.

Rugh, S. E. and Henrik Zinkernagel. "The Quantum Vacuum and the Cosmological Constant Problem." *Studies in History and Philosophy of Modern Physics* 33 (2001): 663–705.

Rundle, Bede. *Why There Is Something Rather than Nothing*. Oxford: Clarendon, 2004.

Sakurai, J. J. *Modern Quantum Mechanics*. New York: Addison-Wesley, 1985.

Schweber, S. S. *QED and the Men Who Made it: Dyson, Feynman, Schwinger, and Tomonaga*. Princeton, NJ: Princeton University Press, 1994.

Shannon, Claude, and Warren Weaver. *The Mathematical Theory of Communication*. Urbana: University of Illinois Press, 1949.

Smolin, Lee. *Three Roads to Quantum Gravity*. New York: Basic Books, 2001.

Steinhardt, Paul J., and Neil Turok. "A Cyclic Model of the Universe." *Science* 296 (2002): 1436–39.

Stenger, Victor J. *Has Science Found God? The Latest Results in the Search for Purpose in the Universe*. Amherst, NY: Prometheus Books, 2003.

———. "Natural Explanations for the Anthropic Coincidences." *Philo* 3, no. 2 (2001): 50–67.

————. *Physics and Psychics: The Search for a World beyond the Senses.* Amherst, NY: Prometheus Books, 1990.

————. *Timeless Reality: Symmetry, Simplicity, and Multiple Universes.* Amherst, NY: Prometheus Books, 2000.

————. *The Unconscious Quantum: Metaphysics in Modern Physics and Cosmology.* Amherst, NY: Prometheus Books, 1995.

Susskind, Leonard. *The Cosmic Landscape: String Theory and the Illusion of Intelligent Design.* New York: Little, Brown, 2006.

Thorne, Kip S. *Black Holes & Time Warps: Einstein's Outrageous Legacy.* New York: Norton, 1994.

Tolman, Richard C. *The Principles of Statistical Mechanics.* London: Lowe & Brydone, 1938. Later editions available from Oxford University Press.

————. *Relativity, Thermodynamics, and Cosmology.* 1934. Mineola, NY: Dover, 1987.

Trusted, Jennifer. *Physics and Metaphysics: Theories of Space and Time.* London: Routledge, 1994.

Tryon, E. P. "Is the Universe a Quantum Fluctuation?" *Nature* 246 (1973): 396–97.

Vilenkin, A. "Birth of Inflationary Universes." *Physical Review* D27 (1983): 2848–55.

————."Boundary Conditions and Quantum Cosmology." *Physical Review* D33 (1986): 3560–69.

————. "Quantum Cosmology and the Initial State of the Universe." *Physical Review* D37 (1988): 888–97.

Weinberg, Steven. "The Cosmological Constant Problem." *Reviews of Modern Physics* 61 (1989).

————. *Gravitation and Cosmology: Principles and Applications of the General Theory of Relativity.* New York: Wiley, 1972.

————. "The Revolution That Didn't Happen." *New York Review of Books,* October 1998.

Wilczek, Frank. "The Cosmic Asymmetry between Matter and Antimatter." *Scientific American* 243, no. 6 (1980): 82–90.

Will, Clifford M. *Was Einstein Right? Putting General Relativity to the Test.* New York: Basic Books, 1986.

Wiltshire, David L. "Viable Exact Model Universe without Dark Energy from Primordial Inflation." http://www.arxiv.org/abs/gr-qc/0503099 (accessed March 23, 2005).

ABOUT THE AUTHOR

Victor J. Stenger grew up in a Catholic working-class neighborhood in Bayonne, New Jersey. His father was a Lithuanian immigrant; his mother, the daughter of Hungarian immigrants. He attended public schools and received a bachelor of science degree in electrical engineering from Newark College of Engineering (now New Jersey Institute of Technology) in 1956. While at NCE he was editor of the student newspaper and received several journalism awards.

Moving to Los Angeles on a Hughes Aircraft Company fellowship, Professor Stenger received a master of science degree in physics from UCLA in 1959 and a PhD in Physics in 1963. He then took a position on the faculty of the University of Hawaii, and retired in Colorado in 2000. His current position is Emeritus Professor of Physics at the University of Hawaii and Adjunct Professor of Philosophy at the University of Colorado.

Professor Stenger has also held visiting positions on the faculties of the University of Heidelberg in Germany and Oxford University in England (twice), and has been a visiting researcher at the Rutherford Laboratory in England; the National Nuclear Physics Laboratory in Frascati, Italy; and the University of Florence in Italy.

Professor Stenger's research career spanned the period of great progress in elementary particle physics that ultimately led to the current *standard model*. He participated in experiments that helped establish the properties of strange particles, quarks, gluons, and neutrinos. He also helped pioneer the emerging fields of very high-energy gamma ray and neutrino astronomy. In his last project before retiring, Professor Stenger collaborated on an experiment in Japan, which showed for the first time that the neutrino has mass.

Professor Stenger has had a parallel career as an author of critically well-received popular-level books that interface between physics and cosmology and philosophy, religion, and pseudoscience.

Professor Stenger maintains a popular Web site, where much of his writing can be found, at http://www.colorado.edu/philosophy/vstenger/.

OTHER PROMETHEUS BOOKS BY VICTOR J. STENGER

Not by Design: The Origin of the Universe (1988)

Physics and Psychics: The Search for a World beyond the Senses (1990)

The Unconscious Quantum: Metaphysics in Modern Physics and Cosmology (1995)

Timeless Reality: Symmetry, Simplicity, and Multiple Universes (2000)

Has Science Found God? The Latest Results in the Search for Purpose in the Universe (2003)

God: The Failed Hypothesis. How Science Shows That God Does Not Exist (in press).

INDEX